# Integrating Water Resources Management

# Integrating Water Resources Management:
## *Interdisciplinary Methodologies and Strategies in Practice*

Edited by
*Geoffrey D. Gooch, Alistair Rieu-Clarke and Per Stålnacke*

IWA Publishing
London • New York

**Published by** **IWA Publishing**
**Alliance House**
**12 Caxton Street**
**London SW1H 0QS, UK**
Telephone: +44 (0)20 7654 5500
Fax: +44 (0)20 7654 5555
Email: publications@iwap.co.uk
Web: www.iwapublishing.com

First published 2010


Typeset in India by OKS Prepress Services.



*British Library Cataloguing in Publication Data*
A CIP catalogue record for this book is available from the British Library

*Library of Congress Cataloging-in-Publication Data*
A catalog record for this book is available from the Library of Congress

ISBN 10: 1843393255
ISBN 13: 9781843393252

# Contents

# Contributors

**Andrew Allan**
IHP-HELP Centre for Water Law,
Policy and Science (under the auspices of UNESCO)
University of Dundee
Dundee, DD1 4HN, UK
Email: a.a.allan@dundee.ac.uk

**Antonio Betâmio de Almeida**
Instituto Superior Técnico, IST
Avenida Rovisco Pais
1049-001 – Lisboa, Portugal
Email: aba@civil.ist.utl.pt

**Line J. Barkved**
Norwegian Institute for Water Research
Gaustadalléen 21
NO-0349 Oslo, Norway
Email: line.barkved@niva.no

**David Barton**
Norwegian Institute for Water Research
Gaustadalléen 21
NO-0349 Oslo, Norway
and NINA- Norwegian Institute for Natural Research
Gaustadalléen 21
NO-0349 Oslo, Norway
Email: david.barton@nina.no

**Dag Berge**
Norwegian Institute for Water Research
Gaustadalléen 21
NO-0349 Oslo, Norway
Email: dag.berge@niva.no

**Fayçal Bouraoui**
Institute for Environment and Sustainability
Joint Research Centre
TP 460, Via E. Fermi, 2749
21027 Ispra (VA), Italy
Email: faycal.bouraoui@jrc.ec.europa.eu

**Dale Campbell**
IHP-HELP Centre for Water Law,
Policy and Science (under the auspices of UNESCO)
University of Dundee
Dundee, DD1 4HN, UK
Email: dalebkk@yahoo.fr

**Johannes Deelstra**
Bioforsk – Norwegian Institute for Agricultural
and Environmental Research
FA Dahls vei 20
NO-1432 Ås, Norway
Email: johannes.deelstra@bioforsk.no

**José M. García-Ruiz**
Instituto Pirenaico de Ecologia, CSIC
Campus de Aula Dei
Apartado 13034
50080-Zaragoza, Spain
Email: humberto@ipe.csic.es

**Anna Maria De Girolamo**
Water Research Institute –
National Research Council (IRSA-CNR)
Via De Blasio 5
70132 Bari, Italy
Email: annamaria.degirolamo@ba.irsa.cnr.it

©2010 IWA Publishing. *Integrating Water Resources Management: Interdisciplinary Methodologies and Strategies in Practice.* Edited by Geoffrey D. Gooch, Alistair Rieu-Clarke and Per Stålnacke. ISBN: 9781843393252. Published by IWA Publishing, London, UK.

**Geoffrey D. Gooch**
Department of Management and Economics
Linköpings University
581 83 Linköping, Sweden
Email: geoffrey.gooch@liu.se
and IHP-HELP Centre for Water Law,
Policy and Science (under the auspices of UNESCO)
University of Dundee
Dundee, DD1 4HN, UK
Email: g.d.gooch@dundee.ac.uk

**Ashvin Gosain**
Indian Institute of Technology Delhi (IITD)
Hauz Khas
New Delhi – 110 016, India
Email: gosain@civil.iitd.ac.in

**Bruna Grizzetti**
Institute for Environment and Sustainability
Joint Research Centre
TP 460, Via E. Fermi, 2749
21027 Ispra (VA), Italy

**Ron Jansen**
IVM-VU,
De Boelelaan 1105
Amsterdam 1081 HV, the Netherlands
Email: ron.janssen@ivm.vu.nl

**K.J. Joy**
SOPPECOM
16 Kale Park, Someshwarwadi Road, Pashan,
Pune 411 008, India
Email: joykjjoy@gmail.com

**Noemi Lana-Renault**
Instituto Pirenaico de Ecologia, CSIC
Campus de Aula Dei
Apartado 13034
50080-Zaragoza, Spain
Email: noemi@ipe.csic.es

**Teodoro Lasanta**
Instituto Pirenaico de Ecologia, CSIC
Campus de Aula Dei
Apartado 13034
50080-Zaragoza, Spain
Email: fm@ipe.csic.es

**N. Latha**
Institute for Social and Economic Change
Nagarabhavi
Bangalore – 560 072, India
Email: latha@isec.ac.in

**Ignacio López-Moreno**
Instituto Pirenaico de Ecologia, CSIC
Campus de Aula Dei
Apartado 13034
50080-Zaragoza, Spain
Email: nlopez@ipe.csic.es

**S. Manasi**
Institute for Social and Economic Change
Nagarabhavi
Bangalore – 560 072, India
Email: manasi@isec.ac.in

**Peter Mollinga**
Center for Development Research (ZEF)
Bonn University
Walter-Flex-Str. 3
D-53113 Bonn, Germany
Email: pmollinga@uni-bonn.de

**Udaya Sekhar Nagothu**
Bioforsk – Norwegian Institute for Agricultural and Environmental Research
FA Dahls vei 20
NO-1432 Ås, Norway
Email: usn@bioforsk.no

**Ingrid Nesheim**
Centre for Development and the Environment
University of Oslo
P.O. Box 1116 Blindern
NO-0317 Oslo, Norway
Email: ingrid.nesheim@sum.uio.no

**Suhas Paranjape**
SOPPECOM
9 Sarvesh, Near Sadguru Gardens,
Thane (East), 400 603, India
Email: suhas.paranjape@gmail.com

**Maria Manuela Portela**
Instituto Superior Técnico, IST
Avenida Rovisco Pais
1049-001 – Lisboa, Portugal
Email: mps@civil.ist.utl.pt

**Antonio Lo Porto**
Water Research Institute – National Research Council (IRSA-CNR)
Via De Blasio 5
70132 Bari, Italy
Email: antonio.loporto@ba.irsa.cnr.it

**Santiago Beguería Portugués**
Consejo Superior de Investigaciones Científicas
Avda. Montañana 1005
50059 – Zaragoza, Spain
Email: sbegueria@eead.csic.es

**K.V. Raju**
Institute for Social and Economic Change
Nagarabhavi
Bangalore – 560 072, India

**Alistair Rieu-Clarke**
IHP-HELP Centre for Water Law, Policy and Science (under the auspices of UNESCO)
University of Dundee
Dundee, DD1 4HN, UK
Email: A.rieuclarke@dundee.ac.uk

**Sergio M. Vicente Serrano**
Instituto Pirenaico de Ecologia, CSIC
Campus de Aula Dei
Apartado 13034
50080-Zaragoza, Spain
Email: svicen@ipe.csic.es

**Eva Skarbøvik**
Bioforsk – Norwegian Institute for Agricultural and Environmental Research
FA Dahls vei 20
NO-1432 Ås, Norway
Email: eva.skarbovik@bioforsk.no

**Per Stålnacke**
Bioforsk – Norwegian Institute for Agricultural and Environmental Research
FA Dahls vei 20
NO-1432 Ås, Norway
Email: per.staalnacke@bioforsk.no

**Haakon Thaulow**
Norwegian Institute for Water Research
Gaustadalléen 21
NO-0349 Oslo, Norway
Email: hth@niva.no

**Anna Madeleine Tiodolf**
Department of Plant and Environmental Sciences
The Norwegian University of Life Sciences
P.O. Box 5003,
NO-1432 Ås, Norway
Email: tiodolf@hotmail.com

# Preface

This book is a major result of the STRIVER research project (htpp://www.striver.no). STRIVER stands for "Strategy and methodology for improved IWRM – An integrated interdisciplinary assessment in four twinning river basins" and was a three year EC funded project running from 2006 to 2009 under the EC 6th framework programme (FP6); contract no. 037141.

In all, thirteen research institutes from nine countries participated in the project and in total, more than 40 senior scientists together with PhD and MSc students contributed to the work. This large group of researchers came from many different academic backgrounds and many different cultures, and 34 of them have also contributed directly to this book in the following 14 chapters. Others, while not listed as authors, have contributed significantly but indirectly through their research. This large pool of scientific knowledge and experience created a unique possibility to explore and analyze Integrated Water Resource Management (IWRM) from various angles and entry points. Our study involved four case basins, the Glomma in Norway, the Tagus in Spain and Portugal, the Tungabhadra in India and the Sesan in Vietnam and Cambodia. Behind the finished chapters in this book lie many hours of lively discussions within the research group, but equally as important for the chapters in this book are the discussions with stakeholders, NGO's and local people in these four basins. These stakeholders included a wide range of interests and expertise, from local fishermen and water managers to policy makers, all of who have interacted with us in very productive and rewarding ways during our field visits and stakeholder meetings. This interaction has been a great source of inspiration when writing this book.

We would therefore like to sincerely thank all these stakeholders and everyone else who has contributed to the project and the book. Thanks also to the Advisory Board for the STRIVER project which consisted of representatives from international water organizations, river basin commissions, water user associations and invited scientists from outside of the project. Thanks also to Dr Susan Baggett who has acted as a secretarial editor for this book volume. Last but not least, the European Commission is gratefully acknowledged for their financial support and we would also like to express a special gratitude to our Scientific Officer at DG Research, Christos Fragakis, who has supported us all along the journey and in encouraged us in our 'striving' for ways to improve IWRM.

7 May, 2010

Geoffrey D. Gooch, Alistair Rieu-Clarke and Per Stålnacke

 *Integrating Water Resources Management: Interdisciplinary Methodologies and Strategies in Practice.* Edited by Geoffrey D. Gooch, Alistair Rieu-Clarke and Per Stålnacke. ISBN: 9781843393252. Published by IWA Publishing, London, UK.

# Chapter 1

# STRIVER in the context of Integrated Water Resources Management

*G.D. Gooch, A. Rieu-Clarke and P. Stålnacke*

## 1.1 IWRM – AN EVOLVING CONCEPT AND ITS BARRIERS

Traditional models of water management, which approach the issue from only one perspective, have come to be challenged more and more and Integrated Water Resource Management (IWRM) is often seen as the way ahead. However, although IWRM principles are widely accepted, a unanimously agreed definition of IWRM has yet to appear. The Technical Advisory Committee of Global Water Partnership (GWP-TEC, 2000) has adopted the following definition that has so far received the most citations "IWRM is a process, which promotes the coordinated development and management of water, land and related resources, in order to maximize the resultant economic and social welfare in an equitable manner without compromising the sustainability of vital ecosystems."

Key points here are *process, coordinated,* and the relationship between *environment/ecosystems* and *economic and social welfare.* IWRM seeks to achieve a balance between economic *efficiency*, social *equity* and *environmental* sustainability. Achieving this balance in a coordinated manner requires consideration of three main components or pillars:

(1) An enabling environment consisting of policies and laws that are centered on IWRM.
(2) Institutional roles at central and local levels, the promotion of public–private partnerships and the recognition that water has to be managed at a river basin level.
(3) Management instruments that will assist institutions in the discharge of their functions. Such instruments are to be used in information gathering, in the assessment of the resource and in developing allocation tools.

The roots of Integrated Water Resources Management (IWRM) can be traced back to the establishment of the Tennessee Valley Authority (TVA) in 1933 (Adams, 2001). The TVA effort targeted many of the elements included in modern IWRM perceptions, i.e., natural resource utilization planning combined with a strategy to balance between social, economic and environmental objectives based on the sustainability concept. The contemporary concept of IWRM was primarily conceived for the purpose of promoting sustainable water management. In 1977, at the International Water Conference in Mar del Plata, emphasis was placed on coordination between water sectors, primarily at a national authority level (Snellen and Schrevel, 2004). The need for sector coordination and holistic approaches was given further attention at the 1992 Dublin conference, preparing for the Earth Summit in Rio de Janeiro. The Dublin statement included the following four principles:

(1) Fresh water is a finite and vulnerable resource, essential to sustain life, development and the environment.
(2) Water development and management should be based on a participatory approach, involving users, planners and policy-makers at all levels.

 *Integrating Water Resources Management: Interdisciplinary Methodologies and Strategies in Practice.* Edited by Geoffrey D. Gooch, Alistair Rieu-Clarke and Per Stålnacke. ISBN: 9781843393252. Published by IWA Publishing, London, UK.

(3) Women play a central part in the provision, management and safeguarding of water.

(4) Water has an economic value in all its competing uses and should be recognized as an economic good.

Later important milestones were the Millennium Development Goals adopted in year 2000, and the Johannesburg Summit on Sustainable Development in year 2002 (http://www.un.org/jsummit/html/basic_info/basicinfo.html). These conferences advocated that IWRM should go further than to just include the co-ordination of sectors and agencies.

In Europe, the Water Framework Directive (EC, 2000) together with the recent Marine Strategy Directive (EC, 2008) provided the legislative and policy structure for operational integrated water resources management, with the aim to protect and enhance the quality of water bodies and to promote a sustainable use of the water resource.

In the 2000s, many scientific publications on IWRM also started to appear. Below we give some examples of the scientific discourse around IWRM and some of its claimed barriers. Gooch and Stålnacke (2006) pointed out that the IWRM definition and contents, which have a strong focus on sustainability thinking, do not, however, provide us with much indication of how this co-ordination, balancing and integration is to be achieved. Other authors have also argued that the IWRM approach, as it is defined by the GWP, cannot be implemented in practice, generally due to operational questions and related problems of establishing measurable criteria (e.g., Jeffrey and Gearey, 2006; Lankford and Cour, 2005; Biswas, 2004). Ballweber (2006) argued that in order to get public and political acceptance for IWRM, the benefits and added-values must be clearly shown. He also claimed that the IWRM *process* is possibly even more important in the short-term than the goal itself. IWRM will ideally include the full range of physical, biological, and socioeconomic variables necessary to manage or develop a region to protect environmental values and provide sustainable human use (Hooper, 2003). Normally, such a knowledge base is lacking or not accessible, or it is scattered across various sector agencies. In addition, social and motivational feasibilities have been mentioned as fundamental attributes for IWRM (Chess and Gibson, 2001), as has the role of science (Gooch and Stånacke, 2010). More specifically, it is often considered that a neutral third party can play a key role, especially in local capacity building (Breen *et al.*, 2004; Ballweber *et al.*, 2005). Others have maintained that IWRM is mainly a political process (Allan, 2007).

The list of scientific publication can be made much longer (e.g., Timmerman *et al.*, 2008; Lenton and Muller 2009; Saravanan *et al.*, 2009, etc). Significnat initiatives have also sought to push IWRM from princple to practice (UNESCO, 2009; GWP/INBO, 2009; UNESCO-HELP, 2009). The point we would like to make and stress is that the overall conceptual IWRM 'thinking', entry-points and approach can differ substantially. Thus there are can be different interpretations and mind-frames of/to IWRM which should be kept in mind.

## 1.2 THE OVERALL AIM AND ROLE OF STRIVER

The intention of this book is to showcase some of the most interesting aspects of IWRM within the four case study basins of STRIVER in relation to the development and adoption of strategies and methodologies. Other publications provide additional insights, see for example, the STRIVER website (www.striver.no), and the edited collection by Gooch and Stålnacke (2010) on the Science-Policy-Stakeholder Interface. However, prior to embarking on a discussoin of the substantive chpaters, the motivation and philoshopy that underwrite STRIVER will be shown. This will hopefully enable the reader to better understand the context in which the research presented in this book was produced.

The European Union Water Initiative (EUWI) (http://www.euwi.net/) launched at the Johannesburg Summit had the overall objective to promote the implementation of IWRM based on a river basin approach and to support – as advocated in the Johannesburg Plan of Implementation – the adoption of water resource management strategies and plans by 2005. As a follow-up to this, the European Commission (EC) via Directorate-General Research launched a 'twinning' mechanism in 6th Framework Programme with the overall goal that it should be a science-based contribution to the EUWI. The aim was to 'twin' European river basins with basins outside Europe to facilitate a platform for the sharing of experiences around IWRM. So far 7 twinning projects have been introduced, one of which is the STRIVER project. In total, the twinning projects cover some 60 case basins in all the five EUWI regions plus South and South-east Asia. It should be noted that the latter regions are not presently covered by the EUWI.

The point of departure for the STRIVER project was the lack of clear methodologies and problems in operationalising IWRM, as pointed out by both scientific and management communities and referenced above. By this is meant the management of the difficult steps between IWRM as a concept, and IWRM in practice. STRIVER therefore had a key aim to use an interdisciplinary approach to both assess and implement IWRM, whilst constantly recognizing the need to engage stakeholders within the research.

Ultimately, STRIVER has sought to advance an understanding of IWRM research and implementation that moves beyond traditional, typically sector-oriented (irrigation, hydropower, water supply, etc.), approaches to water management, and move toward a cross-sectoral approach that focuses much more on the interplay and linkages between the numerous sectors that influence, or are influenced by how water is managed. The general findings of the project are presented in Chapter 14, and specific outputs related to the case study basins are incorporated into Chapters 2–13. However, prior to a moving onto the case study specific issues, this chapter provides a description of the overall approach adopted in STRIVER and the rationale for the choice of case study basins. The intention is to provide an account of how research projects can tackle complex, multi-disciplinary issues as experience by STRIVER.

The complexity of the issues addressed within the STRIVER project meant that the project needed to engage with experts from a wide range of disciplines and backgrounds. The project team was therefore made up of a full spectrum of scientists that represented both natural science and social science disciplines, including law, political science, economics, hydrology and engineering. As important as the need to ensure a range of disciplines was the need to encompass a range of backgrounds and experiences. The project team therefore encompassed universities, private research institutes, governmental organisations, civil society organisations and international policy organisations; from Europe, South Asia and South-east Asia. This representation of a broad range of disciplines and backgrounds was complemented by an advisory board of leading experts and institutions – academics and practitioners in issues related to water resources management. In addition, a strong emphasis on stakeholder interaction, as described in more detail below – supported the objective of ensuring that a wide range of perspectives were taken into account during the design and development of research activities.

## 1.3 THE STRIVER CASE STUDY BASINS

Four twinned river basins that covered six countries in Europe and Asia were selected (Figure 1.1). The case studies, described in greater detail in the individual chapters in this book, were:

(1) the Glomma River basin in Norway;
(2) the Sesan River basin, a tributary of the Mekong, and shared between Cambodia and Vietnam;
(3) the Tagus River Basin, shared between Portugal and Spain; and
(4) the Tungabhadra River basin, which is shared between States within India.

The STRIVER case study basins cover many contexts and issues. The rationale for their choice was not only to capture a wide range of issues, but also to foster synergies between basin experiences through the twinning approach. From a regional perspective the intention in choosing the particular case study basins was to compare approaches to water management adapted within Europe to those undertaken in the developing country context.

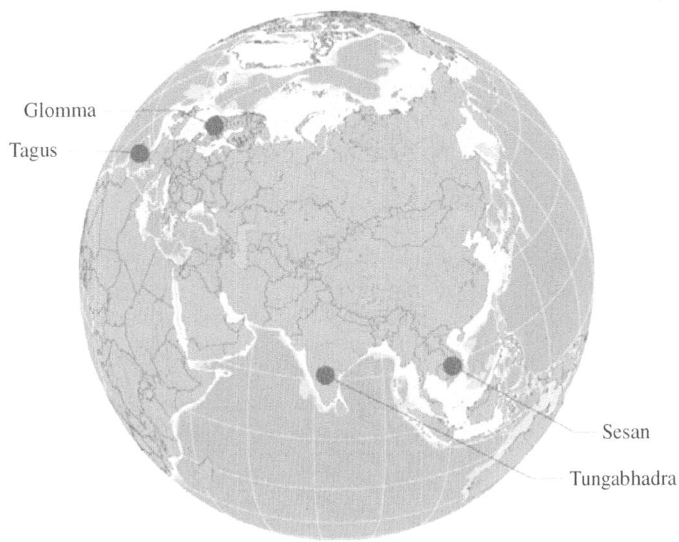


**Figure 1.1** The four case study basins of the STRIVER project: Glomma in Norway; Tagus between Portugal and Spain, Sesan between Portugal and Spain; and the Tungabhadra in India.

More specifically the two European basins of the Glomma and the Tagus provided the opportunity to draw upon the challenges in implementing the EC's cornerstone piece of legislation related to IWRM, namely the EC Water Framework Directive (WFD) (EC, 2000). However, purely from a European perspective, the choice of the Glomma and Tagus basins also provided an opportunity to compare experiences between two very different European contexts; given that in the Tagus water scarcity is a major challenge, whereas pressures on the Glomma largely centre on water pollution.

Hydropower development constituted a major feature in all the river basins, although experiences were very different. For example, in the case of the Glomma there is a long tradition of hydropower agreements and concessions and many conflicts have been successfully resolved. However, some challenges still remain in accommodating the interests of all groups, including energy, fishing, tourism, recreation, agriculture and industry within existing regimes. A similar situation is evident in the case of the Tagus, where hydropower development is predominantly focused on energy production, although as seen in Chapter 7, water for agriculture has become more important due to move from rain-fed to irrigated agriculture. This combination of water for energy and water for agriculture can also be seen in the case of the Tungabhadra River in India, as shown in Chapters 8 to 11. In the case of the Sesan, hydropower projects have a more recent history, and issues around the equable and sustainable operation of existing projects, and the siting and political feasibility of proposed projects remains a pressing issue as shown in Chapter 12. In the Sesan, energy is mostly produced in Vietnam for Vietnamese use while in the Cambodian stretches of the river the people mostly depend on river-side farming and fishery. Claimed impacts from the hydropower development on water quality and subsequent livehoods is another issue as discussed in Chapter 13. Table 1.1 provides an overview of the basins and the key issues that justified their selection for the STRIVER project.

**Table 1.1** Water management issues in the four case study basins (Gooch and Stålnacke, 2010).

| Issues | Tagus | Glomma | Tungabhadra | Sesan |
|---|---|---|---|---|
| Location | Spain/Portugal | Norway | (Karnataka/Andhra Prasesh,) India | Cambodia/Vietnam |
| Drainage area (km$^2$) and Annual precipitation (mm) | 81,000<br>450–1100 | 41,200<br>260–1050 | 71,417<br>450–1200 | 17,000<br>1000–2200 |
| Major water uses in the basin | Hydropower, Agriculture, Domestic use, Irrigation, Tourism, Industry | Hydropower, Fishing, Tourism, Agriculture, Industry | Agriculture, Domestic use, Hydropower | Hydropower, Agriculture, Fishing |
| Major land use (in order of priority) | Agriculture, Urban Settlements | Forestry, Agriculture | Agriculture, Urban settlements, Forestry | Agriculture, Forestry |
| Environmental data, info availability and sharing | OK, but not shared properly especially between the two countries | Good and shared well in relation to the other 3 basins.<br>High public access | Not well monitored and not shared between the two provinces | Poor data collection. Poor sharing within and between the two countries |
| Major problems | Water use and allocation (scarcity), Transboundary conflicts, Pollution Droughts, Fires | Pollution Refined environmental flow assessment needed | Water use and allocation, Water scarcity, Pollution, Transboundary conflicts Deforestration | Transboundary problems Water use and allocation, Lack of pollution assessments |
| Hydropower projects | Major hydropower projects set up on the river Water transfer | Series of 56 hydropower stations and 26 reservoirs | Two major hydropower projects | Major hydropower projects (several planned HPP) |
| Policy | Lack of basin level IWRM policy framework | Well formulated basin level policy | Lack of basin level IWRM policy framework | Lack of basin level IWRM policy framework |

## 1.4 FOCUS AREAS WITHIN STRIVER

The STRIVER project was structured around a number of specific topics that were chosen to promote the widest possible integration amongst several different groups, namely, academics representing different scientific disciplines, water managers, politicians and policy-makers, NGO's and civil society. These topics were then not examined in all case areas.

This section provides an overview of the topic areas that were adopted in the four STRIVER case study basins (Figure 1.2). As noted above, the intention in presenting the ways that these topic areas were examined is to offer an example of how research related to IWRM might be conducted. Specific elements of the topic areas, such as environmental flow, law or modeling are then presented in the subsequent chapters of this book.

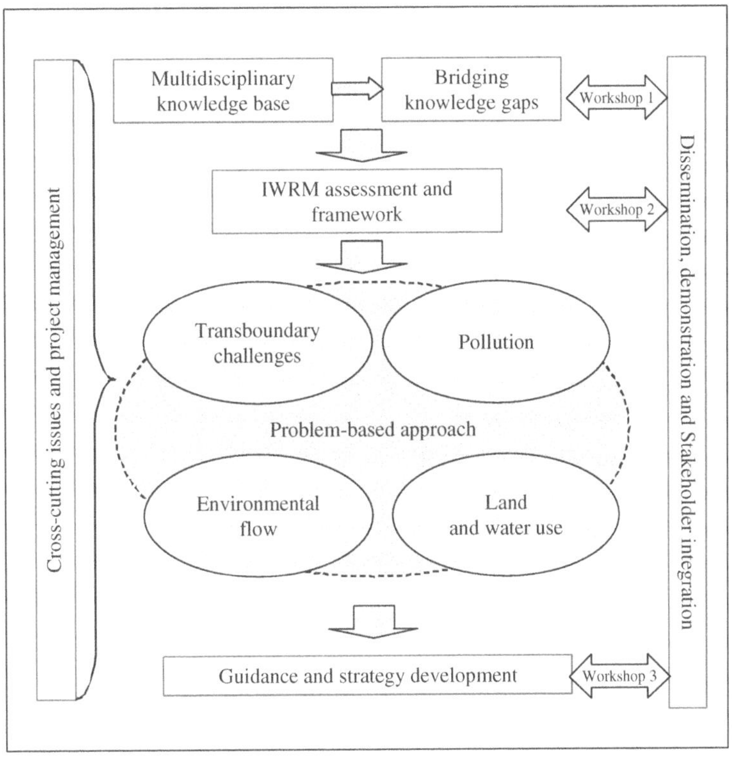


**Figure 1.2** Overview of the focus areas of the STRIVER project.

### 1.4.1 Multidisciplinary baseline assessment and initial stakeholder dialogue

A fundamental prerequisite to the assessment of IWRM at the case study level was to understand the existing status of water resources and water management. The STRIVER project therefore sought to provide a baseline assessment of IWRM within each of the four case study basins. Central to the development of the baseline assessment was the creation of a multi-discipline knowledge base platform which provided an avenue by which to collect data and information related to the status of the river basin regarding hydrology, water quality, user demands, land use, user conflicts, political structures, laws and policies. This data and information gathering exercise also took into account other relevant research concluded under EC FP programmes and elsewhere. While the focus of the data and information gathering was mainly on the case study basins, a review of international literature that was not necessarily directly related to the four river basins was also conducted where it had a relevance to the issues at stake within the basins.

In addition, to gathering data and information from scientific publications, law and policy documents and so forth, an effort was also made during this early phase of the project to engage with stakeholders in order to,

(1) identify who were the key stakeholders within each case study basin, and their relative power, influence and interests;
(2) ascertain synergies with existing projects and initiatives within the four case study basins; and
(3) raise awareness of the STRIVER project, and its aims and objectives.

A first round of stakeholder workshops was organized in each of the case study areas. The selection of the groups was based on information gathered from the stakeholder mapping exercise which was conducted by the project team. These workshops were complemented by a number of smaller focus groups and expert interviews within the case study basins.

### 1.4.2 IWRM assessment

The main output from this initial research was to provide a detailed assessment of the current status of IWRM within each of the case study basins, and also an identification of knowledge gaps where more information would need to be collected. The IWRM assessment involved, examining five components based on the Dublin principles, formulated at the international Conference on Water and Environment in Dublin in 1992, namely,

(1) protection of the catchment and the environment;
(2) measures to ensure efficient and equitable use of water;
(3) institutional analysis and stakeholder involvement in water management;

(4) capacity-building; and
(5) transboundary issues (where and if applicable).

The assessment was aimed not only at assessing to which degree IWRM components could be found in the basins, but also at evaluating the process towards implementing IWRM in the different basins. This was performed by analyzing planning documents in each basin (Nesheim *et al.,* 2008). The research outputs from this topic area are primarily presented in Chapters 2 (Glomma), 5 (Tagus), 8 (Tungabhadra), and 12 (Sesan).

*Protection of the catchment and the environment* involves taking a holistic view and approach to catchment management. This involves regarding water as an economic good while at the same time protecting biodiversity and human needs. Protection of the catchment is central to this volume and to the STRIVER project. *Institutional analysis* is necessary in order to understand the actors and institutions that manage water, as well as the laws and policies that provide the playing field for this management. This analysis has been conducted in all the four Striver basins and recommendations on how to improve institutional efficiency and interaction have been formulated for all four basins. At the center of these recommendations are the insights into the lack of coordination between institutions that can be seen in most of the basins, as well as the need to increase institutional and administrative capacity in many of them. Stakeholder involvement is seen as necessary for IWRM for a number of reasons; first, as a matter of principal, those who are affected by water management decisions should be able to influence those decisions, or at least express their opinions about how water should be managed. Secondly, stakeholder involvement is seen as a way to improve the quality of decisions about water management. Local people, industry, NGO's and civil society have knowledge of conditions in the catchment, knowledge that is different to and complements the knowledge provided by the scientific community and professional water managers. By combining this local knowledge with other forms of knowledge, more informed decisions can be reached. Thirdly, the implementation of efficient water management is dependent on the support of stakeholders. By involving stakeholders in the policy-making process it is more likely that they will support the implementation, the carrying out, of these policies. Institutional analyses and discussions of stakeholder participation are included in the majority of the chapters in this book, and the authors provide innovative insights into how stakeholders can be involved not only in the identification of issues and problems, but also as providers of input into models and scenarios. The ways in which stakeholders have played a vital role in the development of different types of scenarios is one of the most innovative aspects of the STRIVER project.

*Measures to ensure efficient and equitable use of water* are the focus of Chapters 7 and 9, where rain-fed contra irrigated agriculture in the Tagus River (Spain and Portugal) are discussed in Chapter 7 and the use of tanks in the Tungabhadra River in India is discussed in Chapter 9. Other chapters also include discussions of this aspect, for example in the chapters on IWRM in the basins, Chapters 2, 5, 8 and 12. In the Glomma the use of water focuses around the need of the hydroelectrical power plants (HEPs) of water for electricity generation contra the expectations of agriculture and recreational fishing; in the Tagus the discussion concerns the use of water for energy production, water for agricultural and urban use, and water as a recipient of waste water. In the case of the Sesan water is used for HEP production in Vietnam while at the same time it is vital for the poor rural populations of Cambodia who have traditionally depended on river-side farming enabled by the seasonally changes in water levels in the Sesan. In India competition over water is intense, with industry, agriculture and urban needs at the center. The efficient use of water must also, of course, include water necessary to ensure the successful functioning of biodiversity in the rivers, lakes and riparian zones. This environmental flow is specifically discussed in Chapter 4, although it is also included in discussions in some of the other chapters (e.g., Chapter13).

*Capacity building* is an aspect that reoccurs in many of the chapters; capacity-building can refer to the necessity to improve the administrative capacity of water managers in some of the basins, for example in Cambodia this capacity is very low, but it can also refer to the importance of providing stakeholders and the public with the information and knowledge necessary to participate in the making of informed decisions about water management. As can be seen in many of the chapters, in the STRIVER project innovative and creative methods have been used to provide stakeholders with increased capacity while at the same time utilizing the local knowledge that they possess. The direct involvement of stakeholders in modeling, studies of environmental flow etc. are excellent examples of how stakeholder capacity can be increased while at the same time respecting local knowledge.

*Transboundary issues* exist in all the basins except the Glomma, which only has a very small part of its area in a transboundary context, with a very small percent of it in Sweden. The Tagus and Sesan rivers are both truly transboundary in that they flow between two countries, Spain and Portugal in the case of the Tagus, Vietnam and Cambodia in the case of the Sesan. The transboundary context super-imposes extra complications on IWRM as it involves the interaction of legal, political and administrative systems from two different countries. As noted above, these may differ considerably. Take for example the case of the Sesan, flowing between Vietnam and Cambodia. These two countries are vastly different in their levels of economic development and in their political systems. Vietnam, while still a developing country, is developing fast, with a BNP growth rate of 5.3% in 2009, while Cambodia is one of Asians poorest countries. Politically the two countries are also very different; Vietnam is a socialist republic with a centralized and stable political system, while Cambodia struggles with a corrupt and inefficient form of government and administration. Besides these differences, the two countries have a long and bitter history of conflict, with the latest episode, that of the Vietnamese invasion of Cambodia, only a few decades ago. In transboundary IWRM these two countries need to cooperate, exchange information and help to build new futures together. This is a very difficult task, but the STRIVER project was able to make a small contribution to the development of cooperation through the organization of a workshop in Vientiane in Laos at which representatives from Vietnam and Cambodia met to discuss common future plans and scenarios for the river. In the Tungabhadra River in India, while the river is not transboundary in the sense that it flows between two different countries, it is, as noted above, trans-border in the sense that it flows between two quite autonomous states which have the power of legislation and policy over issues concerning water.

## 1.4.3 A problem-based approach

It is important to acknowledge that being too ambitious at the outset, that is, ignoring the political, social and capacity related problems that may hamper an effective project implementation, can result in the formulation of a strategy that is theoretically good, but that does not work well in practice. Experience suggests that major initial reforms are usually not essential to catalyzing change. Instead, an instrumentalist approach is often suitable, where small steps that can easily be implemented is often sufficient to launch the process of moving towards more sustainable water development and management. These observations proved central to the design of the key topic areas of the STRIVER research. It should be remembered, however, that an instrumentalist approach might face difficulties if the necessary changes are large, as institutional inertia may then come into play.

Thus, in order to provide the project with sufficient focus to address key IWRM challenges at the basin level, and through a twinning approach, four key problem-based issues were tackled in detail, namely,

(1) Water regimes in transboundary regulated rivers;
(2) Environmental flow;
(3) Land and water use interaction; and
(4) Water pollution.

These issues were analyzed using a combination of natural science, social science and participatory methods. Stakeholder workshops and other interactions were important means of developing the analysis in each of these four key topic areas. The following sections in this chapter provide a brief overview of this analysis, and references to more detailed information. The main conclusions and recommendations concerning the research methodology and IWRM implementation are presented in Chapter 14.

### *1.4.3.1 Transboundary water regimes in highly regulated rivers (Sesan and Tagus)*

The main objective of this part of the work was to analyze the role of water regimes, actors and institutions in transboundary regulated rivers through Actor-Network Theory (ANT), scenario and governance analysis, and to provide recommendations for integrated policy models for improved water management. A key aspect was to utilize outputs from the baseline assessment and stakeholder meetings in order to identify the major actors in the water regimes, as well as the relationships (multi-level and multi-functional) between them. This was achieved through an ANT approach in which actors (human and natural), and domestic and transboundary relationships are analyzed. ANT is an interdisciplinary approach to the social sciences and technology studies, and closely relates in terms of complexity and locality to the sociology of knowledge and systems theory. Elements of this analysis are presented in Chapter 12 and Chapter 6 of this publication as well as in Gooch and Stålnacke (2010); Gooch and Rieu-Clarke (2010); and Rieu-Clarke *et al.* (2010).

### *1.4.3.2 Water pollution and models (Tungabhadra and Glomma)*

In this part of the project the main objective with the twinning was to analyze the possibilities for mutual transfer of know-how (e.g., experiences, concepts and results) and technology (e.g., methodologies and models) with regard to water pollution. In addition, efforts were made to develop strategies in an integrated framework consisting of different water and land use needs and users for improved methodology for quantifying and analyzing water pollution. The overall methodology includes the use of catchment scale hydrological simulation models, such as SWAT, in the case study basins. Elements of this analysis are presented in Chapters 3 and 11 of this publication, and Grizzetti *et al.* (2010).

### *1.4.3.3 Environmental Flow (Sesan and Glomma)*

The main objective of the twinning related to environmental flow was to improve the methodology for assessing environmental flow in the rivers. There are a number of methods to access environmental flow, but most of them depend on good access to data, something that was not available in the case of the Sesan. In the Glomma River water flow and water level regulations have been in place since 1836. Over the years, new rules have passed, old rules have been adjusted. Currently, a comprehensive set of hydropower regulation schedules for the different sections of the river exists. In the different sections, different motivations, rationales, and methods have been used for assessing these regulation schedules. In Sesan, there are still no adopted hydropower regulation schedules. The STRIVER project therefore sought to apply a new method for assessing environmental flow through adopting simple pressure-impact curves (functions) for the most important ecological and use categories (the Pressure-Impact Curve Method) within the river basins. Research outputs from this analysis are presented in Chapter 4 of this book, and Berge *et al.* (2010).

#### 1.4.3.4 Land and water use interactions (Tungabhadra and Tagus)

The main objective the work around land and water use interactions was to improve the understanding and develop strategies and interventions for sustainable land and water management in an integrated framework consisting of different water and land use needs and users. To reach its objective STRIVER conducted a comparative analysis between the Tungabhadra and Tagus river basins. The research involved a range of activities, including the use of GIS tools, participatory action research methods, land use assessment, livelihood analysis, and efficiency measurements. Element of this analysis are presented in Chapter 10 of this publication, and in Manasi *et al.* (2010).

#### 1.4.3.5 Guidance and strategies for decision support and River Basin management plans in case basins

Evaluation of the final research results in each of the case study basins was a crucial aspect of the STRIVER project. This part of the work utilized the results from the interdisciplinary IWRM assessment in the four case study basins and the twinning work related to water pollution, water regimes in highly regulated rivers, environmental flow and land-water use interactions.

Draft strategies and recommendations were then formulated for each of the case study basins. Through the use of participatory scenarios, as described in Gooch *et al.* (2010), the draft strategies and recommendations were presented to the stakeholder groups during the final project stakeholder workshops. Based on comments received from stakeholders during the workshops, the strategies and recommendations were then modified where necessary, and posted and disseminated in the form of policy and technical briefs (see Chapter 14).

## 1.5 CHAPTER OUTLINE AND APPROACH

As noted above, the purpose of this publication is not to provide an exhaustive list of IWRM assessment methods within all the case study basins, but rather to pick up on some key aspects related to the implementation of IWRM. The book is therefore structured in a way that mirrors the topic areas above. It should be noted, however, that the three main parts of the project have been utilized in all the case areas and topic areas, although all have not been used in all. This means that each part of the project has been based on (a) an initial analysis of existing conditions, (b) original research, either from the perspective of natural social science or economics, or a combination of one or more of theses and (c) interaction with stakeholders, civil society, policy-makers and politicians at all stages of these processes.

After this introductory chapter, which lays out the theoretical and practical frameworks for the project, the book continues with a chapter on IWRM in the Glomma basin, Norway. The Glomma has a long history of HEP development, and also a long history of attempts to regulate the relationships between different water users. As a well-managed basin in a wealthy and politically stable country, Glomma provided the project with an example of what can, or cannot, be achieved with sufficient funding and political good-will. However, it should also be remembered that as the chapter rightly points out, the use of HEP as the engine of industrial (and perhaps democratic) development, should not be underestimated. The move to a more post-industrial society has also created special conflicts, just as the move from agrarian to industrial society did. Today, people's demands on water use have become more focused on recreational purposes, and the use of water for production of HEP that was once necessary to keep the wheels of industry turning has been joined by the demand that water be used in the rivers to provide good opportunities for recreational fishing. This aspect of water use is further developed in Chapter 4, which deals with Environmental Flow (EF) in the Glomma. Chapter 4 also provides insights into how the project built upon Norwegian experiences in the development of participatory methods to establish environmental flow methods, a methodology that was that adapted to conditions in the Sesan River in South-East Asia. This is a very good example of the methodology used in the project, where methods developed in one country were developed to fit in another, under different climatic, social and political conditions.

In Chapter 3 the authors describe their work with another aspect of science-stakeholder interaction, and present the results of their work with participatory watershed modeling. The chapter describes an innovative process for stakeholder involvement, in which stakeholders provide input into model set-up, model validation and finally into model predictions through the use of scenarios. Building on earlier work of Gooch and Stålnacke (2005, 2010) this process of stakeholder participation in future scenarios has been further developed in the Striver project, and used in one way or another in all of the four case basins. The authors conclude that the combination of combining the technical knowledge of the research team with the economic and social concerns of the stakeholders was beneficial at all stages of the process.

In Chapter 4 the authors present a model for involving economic and social values in impact curves which are then used to define environmental flow. The Pressure-Impact Multi-Criteria Environmental Flow Analysis described here represents a significant step forward in the interaction of technical experts and stakeholders, as it builds upon a methodology that allows stakeholders to identify alternatives, and experts to use their expert knowledge of water flows and their effect on the environment and energy production. The possible outcomes of the different alternatives can then be evaluated and rated. The chapter also includes a discussion of the use of computer software, and the relative advantages of custom or generic computer programs. The chapter also discusses the problems of identifying suitable indicators for pressures and impacts and concludes that the involvement of stakeholders has been a positive influence in this work in the Norwegian case area.

In Chapter 5 the focus is moved to the south of Europe, to the Tagus River basin which flows between Spain and Portugal in the Iberian Peninsula. The chapter analyses IWRM in the river basin, with special focus on the transboundary relations between Spain and Portugal, and the legal and institutional arrangements in place. The context of these legal and institutional arrangements is vastly different to the conditions in place in the Glomma, as described in Chapters 2, 3 and 4. While Norway is a single country with a centralized political system, Spain can be described as semi-federal as the regions have a considerable amount of autonomy. Portugal is similar to Norway in that it has a unitary political system. In the case of the Tagus, these two different political and water management systems must interact, and at the same time transboundary water relations in the Tagus are the result of national legal and policy agreements. This of course complicate the development of IWRM in this context, and the authors conclude that intra-basin management and coordination is yet to be fully developed, that agreements need to be enforced more actively, and that stakeholders need to be engaged more actively in IWRM processes in the basin.

Chapter 6 contains an in-depth analysis of the legal framework for the Tagus River, focusing on the complicated interaction of international, national and regional laws and agreements, and on the bi-lateral agreements that have governed water management in the Tagus in the present and past. The Chapter analyzes the need for legal access to information, pointing out that without these formal rights of access, stakeholder participation will be hopelessly restricted. A basic starting point for participation is that people know what the issues are, what the scientific basis of water management is, and which economic, social and political priorities are taken into account. This issue of formal rights to information is also an important aspect of water management in the Tungabhadra River in India and the Sesan River that flows from Vietnam to Cambodia. However, as the authors point out, formal access rights must also be backed up by institutional and political arrangements that actually allow and facilitate this access to information. Finally the authors also stress that access to justice is vital, otherwise existing laws and regulations, how good they might be, may not be implemented.

Chapter 7 continues to analyze conditions in the Tagus River, first presenting a description of climatic and topological conditions in the basin, and then continuing with a description of rain-fed agriculture in the region. While the chapter is predominantly focused on the Spanish part of the basin, similar conditions and challenges can also be found in Portugal. The authors describe how traditional methods and crops were well-suited to the challenging conditions around the Tagus, and how rain-fed agriculture could for the most provide enough food for the local populations. However, the introduction of new crops that need more water, and the implementation of European Union and other regulations such as the Common Agricultural Policy (CA) led to crop failures during years of reduced rainfall. The result of these forces has been the development of irrigation agriculture. The authors conclude that the demands on water from irrigation agriculture, together with growing demands from urban areas need to more efficiently managed through a more realistic use of water concessions.

In Chapter 8 the geographical focus is once again changed, and this and the following three chapters deal with aspects of IWRM in the Tungabhadra River in India. The chapter begins by providing the context of water management in the Tungabhadra. Here, while the context is not transboundary in a national context, it is transborder in that the Tungabhadra flows between two Indian states, each of which is comparable in size and population with many European countries. India also has a federal system of government and jurisdiction over water is not the responsibility of the Union Government. Conflicts across sectors are also common in the basin, and the number of challenges facing water management is considerable and complicated. While the authors stress the need for stakeholder participation in water management in the Tungabhadra, they also claim that this is not happening at the moment and that a participatory approach needs to be developed. They conclude by stressing that legal and institutional changes need to be made at different administrative levels if IWRM is to be successfully implemented.

Chapter 9 continues with the Tungabhadra River and introduces the concept of 'tanks', meaning receptors for water, water that can be gathered during the rainy season and used both during the dry and rainy seasons. The historical development and political use of these tanks by the British is described together with the decline in the use of tanks after

Indian independence in 1947. The reasons for this decline are discussed in the chapter, as is the present situation in the basin. The authors then discuss a case study of four tanks and five villages, and conclude that tanks still play an important part in agriculture in these villages, despite the overall decline in their use. The chapter concludes with some recommendations on how the use of tanks could be combined with other techniques to achieve IWRM.

Chapter 10 continues with the Tungabhadra, this time looking at water quality and water pollution assessment in the basin. The authors describe how they attempted to overcome a lack of data necessary for modeling exercises using SWAT (Soil and Water Assessment Tool) which is a continuous time model at the basin scale designed to predict impacts on water, sediments and agricultural chemicals in basins. While the approach used in this chapter is quite technical, the authors also present the ways in which they engaged local stakeholders. Local meetings were organized and the stakeholders were asked to discuss likely acceptable future changes in the basin. These discussions then formed the basis of future scenarios in a 10 year perspective. The scenarios primarily addressed the following issues, (a) water demand, (b) land use, (c) agricultural practices, (d) waste water treatment practices, (e) sanitation practices and (f) urbanization processes. The scenarios were then used as input into the model. The authors conclude that the use of stakeholder input into scenarios, and from there into modeling, facilitated the formulation of recommendations for best management practices in the basin.

Chapter 11 begins by pointing out the lack of coordination between different actors in the water sector in the Tungabhadra, and continues with recommendation for improved inter-sectoral integration in the basin. The authors then return to the issue of water sources first raised in Chapter 9, and recommend that large-scale (large dams) and small-scale (small tanks etc.) should both be used, without an unnecessary focus on one or the other. The chapter then continues with a discussion which can be found in many of the other chapters in this book, that of the need for institutional changes that facilitate the engagement and participation of many different stakeholders. This is of course dependent on access to data and information and on training and capacity-building, as the authors rightly point out.

In Chapter 12 the focus moves to the Sesan River in South-east Asia. The Sesan flows from the Central Highlands of Vietnam down into the north-eastern provinces of Cambodia. Here the challenges facing IWRM come mostly from the development of HEP dams close to the border in Vietnam. These dams, while necessary for the production of electricity in Vietnam, have cause considerable harm to the poor rural populations living by the Sesan in Cambodia. Communication and coordination between water managers and politicians in Vietnam and Cambodia have been the major challenge to IWRM implementation in the river and the authors present an overview of the problems and a number of recommendations for how to improve the governance of the Sesan.

Chapter 13 continues with the Sesan, this time focusing on water quality and health issues believed to be caused by the HEP dams in Vietnam. The authors present results that demonstrate that skin problems and other health issues in the population living beside the Sesan in Cambodia are probably the result of changes in water flows caused by the construction of the dams in Vietnam. The chapter concludes that while the available data may not allow firm conclusions to be drawn on these issues, there are alarming indications that there are connections, and that more research and comprehensive water quality monitoring should be conducted.

Chapter 14 shows the overall findings and also discusses the overall IWRM implications.

Together the chapters in this book provide a rich and varied picture of IWRM frameworks, of the main issues at stake, of the challenges and problems that face the implementation of IWRM in different parts of the world, and of how one research project, STRIVER has attempted to find solutions to problems and to identify potentials. Throughout the work, stakeholder involvement has been central, as has the interaction between science, policy and stakeholders, which has already been presented in another book which is a product of this innovative and forward-looking project (Gooch and Stålnacke, 2010). It is our hope that the contents of this volume will provide water managers, proponents of IWRM, policy-makers, stakeholder and the public, and our colleagues in the scientific study of water management with insights into the ways in which IWRM, still more a philosophy that a finished method, might be implemented and developed in different parts of the world.

## 1.6 REFERENCES


Adams, W.M. (2001) *Green Development: Environment and Sustainability In The Third World*. 2nd Ed. Routledge, London.

Allan, T. (2007) IWRM: The New Sanctioned Discourse. In *Integrated Water Resources Management: Global Theory, Emerging Practices and Local Needs*. Mollinga, P.P., Dixit, Ajaya and Athukorala, K. (eds.), Sage Publications, New Delhi.

Ballweber, J.A. (2006) A Comparison of IWRM Frameworks: The United States and South Africa. *Journal of Contemporary Water Research & Education*, **135,** 74–79

Berge, *et al.* (2010) The Science-Policy-Stakeholder Interface and Environmental Flow. In *Science, Policy and Stakeholders in Water Management – An Integrated Approach to River Basin Management*. Gooch, G.D. and Stålnacke, P. (eds.) Earthscan, London.

Biswas, A.K. (2004) Integrated water resources management: a reassment. *Water International* **29**(2), 248–256

Breen, C.M., Jaganyi, J.J, van Wilgen, B.W. and van Wyk, E. (2004) Research projects and capacity building. *Water SA* **30**(4), 429–434.

Chess, C. and Gibson, G. (2001) Watersheds are not equal: exploring the feasibility of watershed management. *Journal of the American Water Resources Association* **37**(4), 775–782.

EC (2000) EU Water Framework Directive. (2000) Directive 2000/60/ EC of the European Parliament and the Council of 23 October 2000 establishing a framework for community action in the field of water policy. Official Journal of the European Communities (22.12.2000) L 327, 1.

EC (2008) Directive 2008/56/EC of the European Parliament and of the Council of 17 June 2008 establishing a framework for community action in the field of marine environmental policy (Marine Strategy Framework Directive). Official Journal of the European Union, 25.6.2008, L 164/19–164/40

Gooch, G.D. and Stålnacke, P. (eds.) (2006) *Integrated Transboundary Water Management in Theory and Practice: Experiences from the New EU Eastern Border*. IWA Publishing, London.

Gooch, G.D. and Stålnacke, P. (eds.) (2010) *Science, Policy and Stakeholders in Water Management – An Integrated Approach to River Basin Management*. Earthscan, London.

Gooch, G.D., Allan, A., Rieu-Clarke, A. and Baggett, S. (2010), The Science-Policy-Stakeholder Interface in Sustainable Water Management: Creating Interactive Participatory Scenarios together with Stakeholders. In *Science, Policy and Stakeholders in Water Management – An Integrated Approach to River Basin Management*. Gooch, G.D. and Stålnacke, P. (eds.) Earthscan, London.

Gooch, G.D. and Rieu-Clarke, A. (2010), The Science-Policy-Stakeholder Interface and Transboundary Water Regimes. In *Science, Policy and Stakeholders in Water Management – An Integrated Approach to River Basin Management*. Gooch, G.D. and Stålnacke, P. (eds.) Earthscan, London.

Grizetti, B. Lo Porto, A., Barkved, L.J., Joy, K.J, Paranjape, S., Deelstra, J. Bouraoui, F. and Manasi, S. (2010), The Science-Policy-Stakeholder Interface in Water Pollution Assessment. In *Science, Policy and Stakeholders in Water Management – An Integrated Approach to River Basin Management*. Gooch, G.D. and Stålnacke, P. (eds.) Earthscan, London.

GWP-TEC (Global Water Partnership – Technical Advisory Committee). (2000) Integrated Water Resources Management. TAC Background Papers No. 4. (GWP, Stockholm, Sweden).

GWP & International Network of Basin Organisations, *A Handbook for Integrated Water Resources Mangement in Basins*, http://www.gwptoolbox.org/images/stories/Docs/gwp_inbo%20handbook%20for%20iwrm%20in%20basins_eng.pdf

Hooper, B. P. (2003) Integrated water resources management and river basin governance. *Water Resources Update,* **126**,12–20.

Jeffrey, P. and Gearey, M. (2006) Integrated Water Resources Management: lost on the road from ambition to realisation? *Water Science and Technology* **53**, 1–8.

Koudstaal, R., Rijsberman, F.R. and Savenije, H. (1992) Water and Sustainable Development. International Conference on Water and the Environment. In: Key Note Papers by ICWE Secreteriat, c/o World Meteorological Organisation, Geneva, Switzerland.

Lankford, B.A. and Cour J. (2005) From Integrated to Adaptive: A New Framework for Water Resources Management of River Basins. In *the Proceedings of the East Africa River Basin Management Conference*, Morogoro, Tanzania, 7–9 March 2005.

Lenton, R. and Muller, M. 2009. *Integrated Water Resources Management in Practice – Better Water Management for Development*. Earthscan. 248p.

Manasi, S., Joy, K.J., Paranjape, S., Nagothu, U.S., Campbell, D., Latha, N., Portela, M.M., Betamio de Almeida, A., Machado, M., Raju, K.V. and Portugues, S.G. (2010), The Science-Policy-Stakeholder Interface in Land- and Water – Use Interactions. In Gooch, G.D. and Stålnacke, P. (eds.) (2010) *Science, Policy and Stakeholders in Water Management – An Integrated Approach to River Basin Management*. Earthscan, London.

Nesheim, I., McNeill, D., Stålnacke, P., Sekhar, N.U., Grizzetti, B., Allen, A.A., Barton, D., Beguería-Portugés, S., Berge, D., Bouraoui, F., Campbell, D., Deelstra, J., García-Ruiz, J.M., Gooch, G.D., Joy, K., Lana-Renault, N., Lo Porto, A., Machado, M., Manasi, S., Nhung, D.K., Paranjape, S., Portela, M.M., Rieu-Clarke, A., Saravanan, V.S., Thaulow, H. and Vicente-Serrano, S. (2009) First IWRM assessment report for the four case basins: Glomma, Tagus, Sesan and Tungabhadra. STRIVER Report D5.1, http://www.striver.no

Rieu-Clarke, A., Baggett, S., Campbell, D., Joy, K.J. and Paranjape, S. (2010) The science-policy-stakeholder interface in sustainable water management: creating interactive participatory scenarios together with stakeholders. In *Science, Policy and Stakeholders in Water Management – An Integrated Approach to River Basin Management*. Gooch, G.D. and Stålnacke, P. (eds.) Earthscan, London.

Saravanan, V.S., McDonald, G.T. and Mollinga, P.P. (2009) Critical review of integrated water resources management: moving beyond polarised discourse. *Natural Resources Forum* **33**, 76–86.

Snellen, W.B. and Schrevel, A. (2004) IWRM: for sustainable use of water – 50 years of experience with the concept of integrated water management. Background document to FAO/Netherlands Conference on Water for Food and Ecosystems, http://www.fao.org/ag/wfe2005/docs/IWRM_Background.pdf

Timmerman, J.G., Pahl-Wostl, C. and Moltgen J. (2008) *The Adaptiveness of IWRM – Analysing European IWRM Research*, IWA Publishing, London.

UNESCO (2008) *IWRM Guidelines at the River Basin Level*, http://www.hydrology.nl/ihppublications/169-iwrm-guidelines-at-river-basin-level.html

UNESCO-HELP (2010) *Hydrology for the Environment, Life and Policy*, http://www.unesco.org/water/ihp/help

# Chapter 2

# IWRM status in the Glomma River basin

*I. Nesheim, P. Stålnacke, U.S. Nagothu, E. Skarbøvik, L.J. Barkved and H. Thaulow*

## 2.1 INTRODUCTION

The challenges in the catchment area of the Glomma River are quite typical of the integrated water management challenges in other catchments in Norway. These are related to the impacts of hydropower development; pollution from agriculture, urban settlements and industry. At the same time the history of the Glomma River reflects Norway's increasing efforts to involve the public in hearing processes in order to ensure that stakeholder interests are taken into account.

In Norway, rivers have been used as a source of power for centuries. Written sources refer to grain mills that were powered from water wheels as early as the beginning of the 11th century (Eie *et al.*, 1996). The view that "water is power" is reflected in a letter from King Håkon V in 1312, where he asks his son-in-law to build a water mill at Borregård (the Glomma basin) (Eie *et al.*, 1996). Since then, technological development has enabled the use of power derived from water for many other purposes, and hydro-electricity became fundamental to the economic development of the country (Hveding, 1992). The first hydroelectric power station was built in 1877 and industry started to use electricity in 1885. From about 1900 to 1940, around 2,000 hydroelectric power stations were built (Hveding, 1992). This development continued at a rapid pace, and in 1990, hydroelectricity production was four times higher than in the 1960s – although there was a stagnant period with regard to hydropower development during the 2nd World War. For some, a waterfall was not appreciated for its beauty or its important ecosystem value, but merely in relation to its power potential. In the Glomma River basin, with a relatively high population density in a Norwegian context and increasing industrial activities, including among others pulp and paper mills, there was a great demand for power. In addition, sewage and waste water from households and industry increased over the years, while more intensified agricultural activities led to higher runoff of nutrients. Disappearing waterfalls, reduced fish catches and increased pollution of waters reduced the recreational value of nature, and this finally awakened people's engagement in the 1970's.

In this context of water for hydropower, and a history of – in some places – severe pollution levels, the chapter seeks to assess the status of integrated water resources management (IWRM) using the Glomma River basin as a case study. The Glomma River is an interesting case because of the intensity and great variety of activities undertaken in the basin, including both industrial and recreational activities. This chapter will assess to what degree IWRM components can be found in the basin, and also evaluate the process towards implementing IWRM. The assessment of IWRM in the Glomma basin focuses primarily on the IWRM components described in Chapter 1 of this book, namely,

(1) protection of the catchment and the environment;
(2) measures to ensure efficient and equitable use of water;
(3) institutional analysis and stakeholder involvement in water management;

©2010 IWA Publishing. *Integrating Water Resources Management: Interdisciplinary Methodologies and Strategies in Practice.* Edited by Geoffrey D. Gooch, Alistair Rieu-Clarke and Per Stålnacke. ISBN: 9781843393252. Published by IWA Publishing, London, UK.

(4) capacity-building; and
(5) transboundary issues.

With regard to the component *measures to ensure efficient and equitable use of water*, we focus primarily on measures to ensure water as an economic good, as there has, in general, always been a sufficient amount of water in this basin – with the only exception being in areas lying in the rain shadow of the western mountain ranges. The IWRM components of *capacity building*, and in particular *transboundary issues*, are given less coverage due to their relatively low relevance in the Glomma basin.

## 2.2 THE GLOMMA RIVER BASIN AND ITS PRESSURES

The Glomma River is Norway's largest river; its catchment area covers 42,000 km$^2$, or 13% of Norway's total land area (Figure 2.1), and includes the Glomma and the Laagen Rivers. The annual precipitation in the area ranges from 260 to 1050 mm. The north-western parts consist of high mountain areas, including the Jotunheimen area with Norway's highest peak, Galdhøpiggen (2468 m asl.), and Rondane and Dovre National Parks. These areas are sparsely populated with a relatively high density of regulated rivers. The eastern parts are covered by forest, whereas the more populated central and southern parts comprise large agricultural areas by Norwegian standards. In total the agricultural areas cover 5.8% of the catchment. Though the population density in the basin is only 29 persons/km$^2$, the basin is the most populated river basin in Norway with a total population of about 675,000.

Historically the Glomma River has been important for the local population with regard to nearly all aspects of life: food production, drinking water, transport and energy; and to varying degrees these user interests still prevail. However, hydropower development has been the most dominant interest, leading to a heavily regulated river (Table 2.1). Indeed, the Glomma River basin currently has 57 hydropower stations, and 26 reservoirs and diversions schemes (Figure 2.1). Despite this regulation, the impact on the river ecosystems has been less than might be expected. Most of the hydropower reservoirs were originally natural lakes and there are few artificial water bodies in the catchment. Many of the other power schemes are run-of-the-river regulations with no storage capacity and only a small intake dam to secure a steady flow of water. As these run-of-the-river regulations hardly influence the water flow in the downstream reaches, there is no need for environmental flow in these rivers (Berge *et al.*, 2008). Nevertheless, the ecological continuum; an important concept in the Water Framework Directive (WFD), is broken because of the dams, and many hydropower projects have affected fisheries negatively, as exemplified by the 150 km long river stretch from Høyegga to Rena, where 80% of the water is diverted to Rendalen, creating a large impact on fish yields in the river (Figure 2.2). Though the larger reservoirs have primarily been located in the remote highland areas, with no or very few human settlements, these higher situated reservoirs are often more regulated, such as the Lake Raudalsvatn (in the Upper Otta regulation) with 30 meters between maximum and minimum reservoir water levels which reduces fish productivity quite seriously (Berge *et al.*, 2008). Increased attention has thus been given to mitigation efforts in order to improve recreational fishing; introducing fish stocking, fish ladders, habitat mitigation and rehabilitation (e.g., Østdahl *et al.*, 2002; Saltveit, 2006).

**Table 2.1** The main drivers-pressures-impacts in the Glomma river basin.

| | Drivers | Main pressures | Main impacts |
|---|---|---|---|
| Glomma | Hydropower | Changed hydrological conditions. Break of ecological continuum. Heavy modification of water bodies (HMWB). | Change in biodiversity due to habitat changes (of which the conditions for salmonid fish species have been particularly focused) |
| | Agriculture | Pollution (fertilisers and to some extent pesticides) added to the river system from mainly diffuse sources | Eutrophication of (mainly) the downstream parts of the river basin, including Lake Mjøsa |
| | Urban areas | Pollution (sewage, runoff, industry) as well as hydro-morphologically modified water bodies | Pollution of the downstream parts of the river basin, including Lake Mjøsa (eutrophication); as well as HMWBs due to urbanisation |
| | Industry | Introduction of pollutants such as environmental toxins, nutrients (point sources) | Toxins in food chains, eutrophication |
| | Long-range transported substances | Increased levels of sulphur and NOx in the waters | Acidification of waters with subsequent ecological problems (e.g., fish death) and need for measures (liming) |
| | Introduced species | Introduced fish and mussel species | Competes with economically interesting species (e.g., minnow competes with young trout). |

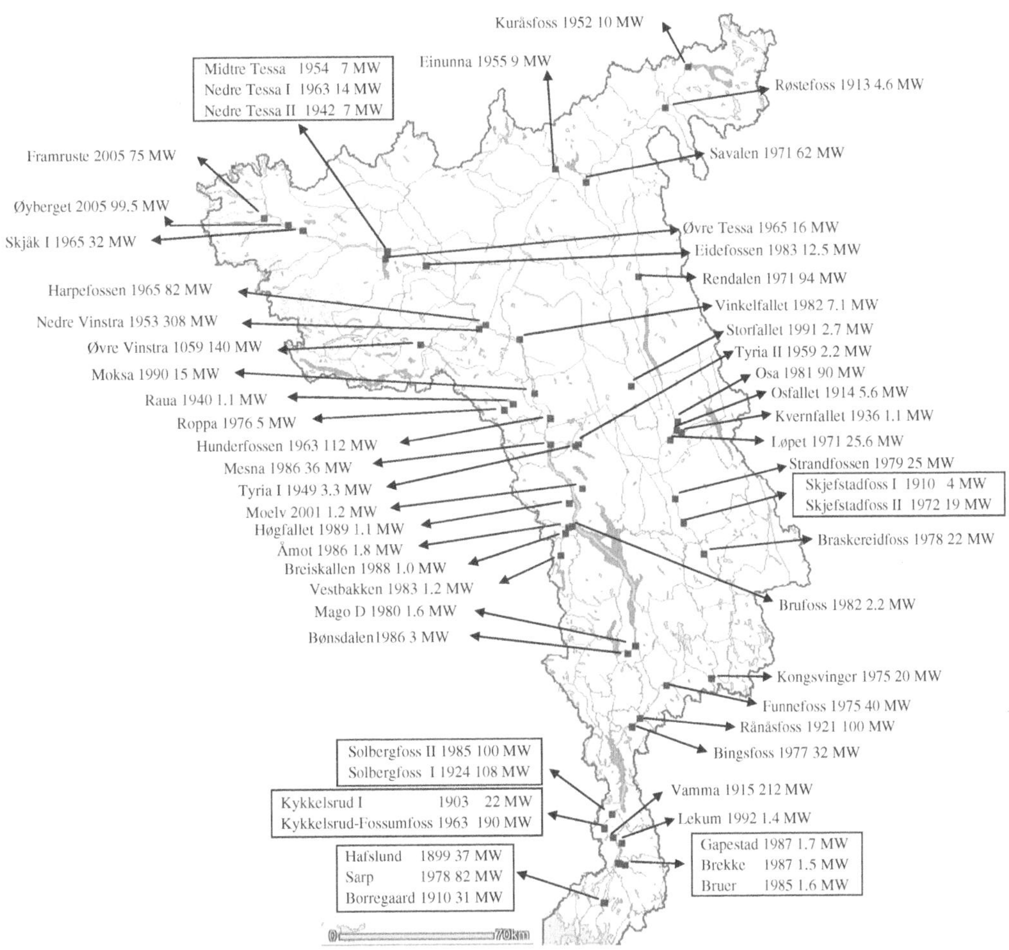


**Figure 2.1** Map showing the Glomma river basin in Norway and the various reservoirs, diversions and power stations.

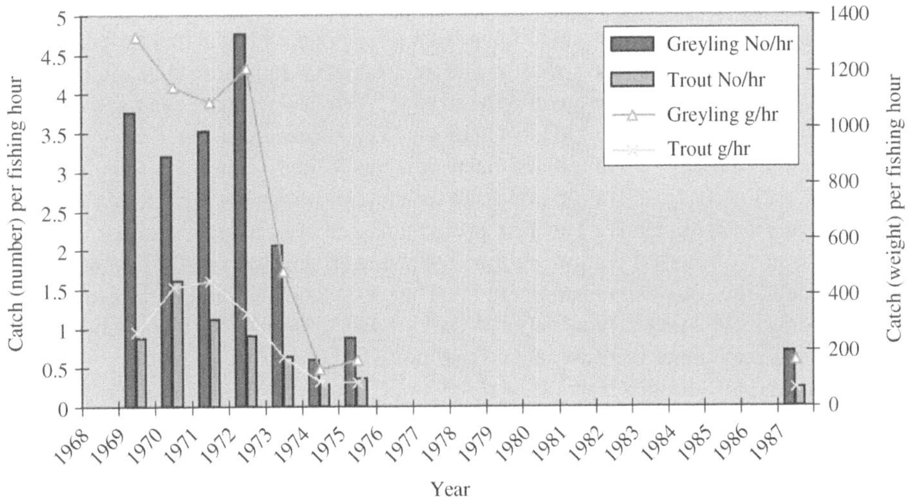


**Figure 2.2** Fish yield in River Glomma in Storelvdal before and after diversion of 80% of the water to River Rena in 1973 measured as catch per unit effort (CPUE) (Quenild and Linløkken, 1989).

The Glomma River is also used as a recipient for point source emissions from municipal and industrial wastewater, and some settlements without efficient effluent treatment still exist (Fazi *et al.*, 2009). Agriculture is another pressure. This is further described in the following section. Other problems are:

(i) environmental toxins entering the food chain, making it risky to consume fish in some areas;
(ii) introduction of new species, and
(iii) long-range transport of air-borne pollutants (Table 2.1).

Of these, the cause of the latter cannot be dealt with within the river basin, but mitigation through adding calcium to the river courses is quite common both in the Glomma and other affected Norwegian rivers. Along the river, agriculture uses water for irrigation, but not enough to affect the water flow notably.

An increasingly larger part of the economic interests are linked to tourism and recreation, comprising both winter sports such as alpine and cross country skiing and summer activities such as hiking, hunting and recreational fishing. These activities need a clean and undisturbed nature to be attractive, a challenge both for the hydropower sector, the municipal sewage treatment plants and the industrial discharge management.

## 2.3 IWRM STATUS IN GLOMMA

The comparative IWRM assessment presented in this chapter is based on two basin reports; one that assesses the general performance and status in relation to seven selected IWRM principles, and another that reviews the implementation of the selected principles by analyzing planning documents in the basin (see Nesheim *et al.*, 2009). At the outset it should be noted that these planning documents serve only as examples of the implementation of IWRM.

### 2.3.1 Protection of the catchment

Environmental pressures in the Glomma River basin are mainly related to hydropower production, agriculture, and industry. Although these pressures still exist, it is interesting to note that the most severe conflicts have occurred in the past and that they have now to a large extent been dealt with.

In Norway, several legal and administrative instruments (such as laws, regulations, licensing practices and also international agreements (see the section on institutional analysis) exist for the protection of the environment and water. The present protection instruments are quite advanced compared to the instruments available during the first half of the 19th century. Hydropower was the first main challenge in the Glomma River system affecting the hydrology of the rivers, biodiversity, the fishery and the aesthetic aspects. The first initiative to protect Norwegian river basins against hydropower development involved a country wide plan for the protection of water courses against hydropower development in the 1960s (Eie *et al.*, 1996). The first overview of potentially protected areas was developed by the so-called "Gabrielsen Committee" in 1963. This work has later become a basis for future protection plans.

Although hydropower at the time was seen as a symbol of economic growth and development, the government stated in the first country wide plan that some of the economic value in not yet developed river courses should be sacrificed for the benefit of environmental interests. Irrespective of this intention however, the intensity of hydropower development continued and conflicts between environmental organizations and concession authorities led to extensive confrontations. License applications were in the 1960s and 1970s granted almost without exception, and if this level of growth had continued all remaining hydropower resources would have been developed within a 15 year period. It was evident that there was a need to differentiate between rivers that should be protected and those that might be developed; and to develop a national tool to prioritize between projects. (Eie *et al.*, 1996). Two master plan activities were initiated: the Watercourse Protection Plans in 1970 and the "Master Plan for Water Resources" in 1980 (WCD, 2000).

A total of five Watercourse Protection Plans have been developed, and within the Glomma River basin, a total 27 smaller watersheds are protected against hydropower development; an area which comprises approximately 30% of the total river basin area (Østdahl, 2002). The first protection plan was launched in 1973 and the last in 2009. 341 Norwegian Rivers are now protected against further hydropower development. The Glomma River basin also comprises three national parks, the Jotunheimen National Park, the Rondane National Park, and the Dovrefjell National Park established in the period 1962–1980. The national parks have a strong legal protection against development projects or other impacts that may affect environmental conditions.

Following its initiation in 1980, the first version of the Master Plan for Water Resources was presented in 1984. The plan included mandatory impact analyses for a large number of projects, taking into account a number of user interests in the water course. Potential hydropower projects were ranked along a continuum from those with low conflict to those with high conflict levels. Projects characterized as having low conflict levels were to be considered for development first, while those with higher conflict levels were to be preserved for other uses than hydropower. In this process, it was considered particularly important to ensure that all affected parties were given the opportunity to read the reports and submit their comments. The plan's main objective was to achieve a nationwide collective management of river basins, based on a set of economic, social and environmental considerations. The plan was important for reducing conflict levels, as well as for ensuring that the most vulnerable and important water courses were protected from development. The plan was also significant in that it was the first initiative aiming at an *integrated management* of rivers throughout Norway.

Minimum water flows are suggested for each project in the Master Plan for Water Resources. In general minimum flow requirements are found in the Water Resources Act (2000). However the majority of the regulations in the Glomma river were developed before 2000. In some of the older regulations, such as the Vinstra regulation from 1953–59, there is no minimum water release and the river is therefore completely dry downstream of the reservoir. It is however, expected that with the renewal of the concession which will take place in the near future, requirements for minimum water flow will be issued. In the Glomma River basin with its many reservoirs and power plants it is of crucial importance that all regulations, operating schemes and mitigation efforts, including minimum flow requirements, are co-ordinated in an integrated watershed management approach (Østdahl *et al.*, 2002).

Since the early 2000s there has been a shift in focus, from the development of large hydropower schemes to the development of small-scale hydropower plants, and on upgrading and modifying already existing schemes. However, although small-scale hydropower projects (power plants with an installed capacity of up to 10 MW) often are cited as environmentally friendly alternatives for energy production, they may still have negative environmental impact. Often a weir or small dam is built in a stream and the water is led through pipes to a small power plant. Except during floods, there will be little or no water left in the stream, and the weir will also act as a barrier for fish and other species. A few such power plants are perhaps not enough to provide major environmental drawbacks, but the combined effect of many such small plants has not yet been properly assessed. Presently there are hundreds of applications to develop such small plants, and it is therefore important to consider the overall impacts. Moreover, a renewed focus on the development of alternative energy sources, including hydropower, to abate carbon emissions to the atmosphere may result in a more positive focus on larger-scale hydropower development in Norway.

Agricultural production, which has an impact on pollution levels, is another pressure in the basin. About 40 years ago, pollution levels IN THE RIVER increased alarmingly and many parts of the Glomma River basin were exempted from recreational uses. Lake Mjøsa, Norway's largest lake, came close to being locked in an irreversible trend of diminishing water quality. During the last 10–15 years, however, the impact of agriculture on water quality has been reduced, although conflicts between water users still occur. The Glomma River receives discharges from municipal wastewater treatment plants and industry, and though the wastewater is, with very few exceptions, treated, there are still many small settlements without efficient effluent treatments. This, along with agriculture and leaks from old industrial deposits, still reduce water quality, creating conflicts for several hundred thousand people who use the river as a drinking water supply. The most important factors contributing to the reduced pollution levels in the river have been the international agreement on reduction of pollutants to the North Sea (The North Sea Agreement), and the Lake Mjøsa Campaign, both having vital importance for the water quality improvements in the basin.

#### 2.3.1.1 The successful Lake Mjøsa abatement programme

The most important water quality issue in the Glomma River basin was the excessive eutrophication of Lake Mjøsa in the 1960s and 1970s due to nearly 100 years of neglect regarding measures to deal with sewage, agricultural and industrial pollution. At that time, about 200,000 people suffered from bad quality drinking water. The eutrophication was felt by everyone, for instance, rocks on the beaches were slimy and slippery and the lake had a foul smell. In 1970–72 the first alarm was sounded: the ecological balance was changing, and new dominant algal species indicated a strong degree of polluted water. If the source of the pollution was not reduced quickly, the lake would go into an irreversible trend of diminishing water quality with regular algal blooms, and Mjøsa as a recreational area would be ruined. During the summer of 1976 the situation worsened, and forced the Minister of Environment in Norway at the time, Gro Harlem Brundtland to launch the 'Action Lake Mjøsa' campaign. The campaign aimed at reducing the phosphorus load to sub-critical levels of 175 tonnes per year. The Mjøsa Campaign found wholehearted support both locally and nationally, not least because of very active media coverage. Newspapers, radio and TV publicized all kinds of stories relating to the campaign. The local newspaper, Hamar Arbeiderblad, introduced the idea of collective action by the housewives in the Lake Mjøsa district to save the lake. The housewives were to stop using detergents containing phosphates; and within a year, a voluntary agreement to reduce phosphates was made between the government and the detergent producers. Grassroots mobilization supported the campaign leaders from many neighborhoods and organizations. All pollution sources were studied, measures were implemented, and responsible persons were identified by name. By 1980 the phosphorous load had been reduced from 459 tonnes per year to 225 tonnes. The set goal of 175 tonnes was not reached before the campaign ended, but Lake Mjøsa was saved. In short, a wide variety of actions were behind the success, political, technical, scientific and public awareness building. Perhaps the most important element was money. Enough money to run the operation was provided, the equivalent of NOK 1225 million at 1977 prices (Ministry of the Environment, 1979). The immediate problems of lake Mjøsa were solved, but some pollution issues still remain in many tributaries. Climate change has increased the temperature in the lake, and more measures have to be implemented to keep Mjøsa and its tributaries clean, and achieve the goal of the WFD, i.e., good ecological status.

#### *2.3.1.2 The EU Water Framework Directive (WFD)*

The most important recent measure to protect Norwegian river basins is the implementation of the EU Water Framework Directive (EC, 2000). The Directive came into force for all EU member states on 22nd December 2000, and was ratified in Norway in December 2006. However, work related to the WFD started early, including a first characterization of water bodies throughout the country, as well as the development of guidelines for certain issues of particular importance in Norwegian rivers, such as guidelines for water bodies heavily modified by hydropower developments (NVE, 2004). In addition 30 river basins have been selected to follow the EU deadlines, of which two sub-catchments are situated in Glomma River. The implementation of the WFD in the remaining 120 Norwegian sub-basins will be carried out from 2010 onwards.

The main objective of WFD is "good ecological status" for all water bodies. This goal comprises requirements for the ecological, chemical and hydro-morphological status of the water bodies. If a water body is characterized as being at risk of not achieving the main objective of good environmental status, measures must be undertaken in order to reverse the negative development. The implementation of this directive has huge consequences on Norwegian water management for several reasons:

(i) the requirements are quite strict; as the water quality in any water body should not differ much from natural or reference conditions;
(ii) the organization of the Norwegian water management has been modified due to the directive;
(iii) there is increased demand for monitoring, especially biological monitoring, which increases the knowledge on the environmental state of the waters; and
(iv) there is increased demand for implementing measures in order to improve the environmental state of the water bodies.

Exceptions from the agreed objective of good ecological status comprise those water bodies that are designated as heavily modified water bodies (HMWB). These water bodies do not need to achieve good ecological status if this means that the changes of their hydromorphological characteristics will have adverse effects on, *inter alia*, "water regulation, flood protection, land drainage, or other equally important sustainable human development activities". In other words, if returning a HMWB to good ecological status will mean that the production of electricity will be seriously reduced, then the goal of that water body changes into "good ecological potential". There has been much work done in Europe to define this goal by various pan-European coordination groups entitled "Common Implementation Strategy" (http://ec.europa.eu/environment/water/water-framework/objectives/implementation_en.htm). In Norway a relatively pragmatic line seems to have been adopted, where the costs and effects of the mitigation efforts are rated against the loss of power production (e.g. Skarbøvik *et al.*, 2006). One interesting issue arises however: the cost of releasing a minimum flow is often not rated as being too expensive (or resulting in a too high a loss of power production); hence this is a mitigation measure that may be issued in most or all rivers downstream of dams. However, these requirements may in some cases be in conflict with the current concessions. Since these concessions often are given for several decades by the highest authority in Norway, the King in Council, such conflicts are not readily solved. A pragmatic solution is to implement "good ecological potential" in the water bodies whenever a concession comes up for renewal, but strictly speaking this is not an implementation practice that is according to the requirements of the WFD. Also, this will not ensure the integrated catchment approach that the WFD requires.

Another interesting issue is that in Norway, measures to mitigate the adverse impacts from hydropower production have to a large extent focused on salmon. The release of water from reservoirs has therefore often been timed to ensure proper water flow during the seasons of migration, spawning, etc. This may, however, mean that water levels in the reservoirs are lowered at a time that is unsuitable for the spawning of other, less economically interesting fish (char and arctic char). The WFD does not stipulate that some species are more important than others; it rather strives to ensure that the original biodiversity is intact. This also means that a typical Norwegian measure to introduce trout from fish farms into the reservoirs, when trout did not originally occur there, is against the regulations of the Directive. Furthermore, the Directive is not concerned with human needs for aesthetics and many of the weirs that are constructed in order to ensure that water covers the entire width of a former river bed may go against the requirements of the Directive, especially if they favor introduced species (e.g. minnow; *Phoxinus phoxinus*). The work on the WFD has called for a new water management structure, as will be outlined below.

### 2.3.2 Measures to ensure efficient use of water

The significance of "efficient use of water" is obvious in those major regions of the world where water demand exceeds water supply precipitation (including potential evapotranspiration) (Stanhill, 1986). However, in some countries such as Norway, where there is in general a surplus of water, the significance of "efficient use of water" has less relevance. But even if there is no water deficit on an annual basis, water may be limited on a short term seasonal basis.

The use of water for hydropower production, or for industrial purposes, can have a large impact on other users' possibility to use water, and conflict of interests may arise. The optimum level of water conservation and use by different users may have to be decided by the government and imposed by legislation, with costs allocated among the various water users in order to ensure efficient and equitable use of water.

In order to allocate costs among the different water users, there is a need to analyze the costs involved in the provision of water, and in the use of water. These include not only the direct costs – of providing a reliable, accessible and good quality supply – but also the "opportunity cost" of water, that is, the costs associated with depriving society or other sectors of the possibility to use this water (McNeill, 1998; Rogers *et al.,* 1998).

As already described in the sections above, hydropower production has had an enormous impact on Norway's economic development, being the key to its industrialization. Not only does hydropower currently provide 99% of the country's electricity, quite a lot is also exported, providing income to the state, the municipality and of course to the hydropower companies. Power generation is, in many municipalities, the most important economic activity. And like all other businesses, the municipality has the right to tax these activities. The municipalities receive income through natural resource taxes and property taxes, while the state also receives electricity tax and valid added tax (VAT). In addition, the concessionaire has to pay concession fees to the affected municipality for permission to use waterfall rights and to regulate the water course. The concession fee is partly intended as a share of value added to the affected district and partly a compensation for general damages and disadvantages. The hydropower companies have, since the first licensing laws from the early 1900s, also been obliged to cede a share of power production, the so called "concession power", to the affected district. Parliament's original purpose with this decision was to ensure that the district that ceded their natural resources to cities and central areas with electricity and industry would benefit from economic development.

Households and farmers and other users also have to pay according to the quantity of water being used, as well as for sanitation. The pricing of water for households however, follows the principle that fees should not exceed the municipality's costs related to the water and sanitation sector. Annual fees shall be calculated on the basis of measured or estimated water consumption. Higher or lower annual fees can be imposed if water or wastewater of larger units involves significantly higher or lower costs (Regulations relating to municipal water and wastewater fees, 2004). Different fees can also be established for sewage-related pollution that deviates from common municipal wastewater. In addition, the principle is that the polluter should pay, as expressed in the pollution law (Forurensningsloven, 1981) states that the municipality should cover all the costs of sewerage sector through fees. Despite the existence of charges and taxes such as these, it may be debated whether, in Norway, the management of water resources has included the perception that water is an economic good. Fresh water has been priced low and has been considered a free good.

### 2.3.3 Institutional analysis

#### *2.3.3.1 The water governance system in the Glomma and the integration principle*

The responsibility for water management in Norway involves several governmental institutions, each with its respective legislation. The Ministry of Environment, with the Directorate for Nature Management and the County Administration as key actors, have the overall responsibility for the implementation of the WFD, while the Norwegian Climate and Pollution Agency is responsible for water pollution issues. The Ministry of Petroleum and Energy, with Norwegian Water Resources and Energy Directorate (NVE) as the key executive institution, has the overall responsibility for managing water as a resource (water quantity). NVE has the legislative power according to the Energy Act (Energiloven, 1991) and the Water Resources Act (Vannressursloven, 2000) to issue regulations and to make individual decisions and prepare cases to be resolved by the Ministry of Petroleum and Energy. Tasks concerning the environment at regional level, on the other hand, are delegated to the County, while the municipalities have, through the Planning and Building Act (Plan og bygningsloven, 1985) the central responsibility for land use planning along river banks. Local municipalities have in addition been delegated the authority to decide on local water pollution issues. Recently, responsibilities have been increasingly delegated to municipalities due to the WFD.

A committee responsible for the coordinated implementation of the WFD in Norway has been established with representatives from nine directorates in total. The committee advises the ministries and coordinates the Norwegian participation in the European Common Implementation Strategy (CIS). Norway has according to the WFD been divided into 11 river basin districts (or 16 if the international basin districts are counted). The county is the competent authority in each respective river basin district with the responsibility to develop an overall regional management plan based on the work performed in river basin committees which consists of municipalities and private actors. The implementation of the WFD in the Glomma River Basin District is governed by the County Administration of Østfold (since 2010). The Department of the Environment at the County Governor plays a central role, and is part of the river basin district (RBD) Committee, which consists of authorities at the regional and local level. In addition, a reference group consisting of various public and private stakeholders advises the process.

The operation and management of the water resources also includes the participation of non-governmental organizations (NGOs) and professional associations. As such, a national reference group for the WFD has been established, comprising 11 national organizations representing various branch organizations, user interests and the Norwegian Association for Local and Regional Authorities. In the River Glomma, an important professional stakeholder, operating at basin level is the Water Management Association (WMA), "Glomma and Laagen Brukeseierforening" (GLB). The GLB is a cooperative body of all hydropower producers in Glomma, with the responsibility to coordinate the use of water in the river for the common benefit of all power plants and to operate reservoirs and regulate the flow in rivers. To maintain the status of a cooperative body for the common benefit of all members, the GLB can neither own power stations nor produce hydropower, but it may own the water facilities and has the authority to regulate. Besides the GLB, there are a number of other NGOs working in Norway on environment and conservation, agriculture, hunting and fishing issues.

### *2.3.3.2 Institutional Integration*

In Norway responsibility for water management can be found at national, regional and local levels and includes ministries, counties and municipalities (Figure 2.3). Further, responsibilities for water management are divided between water resources management (quantity) and water quality management (quality). As has been described above, a split exists between the legal and institutional arrangements for qualitative and quantitative aspects of water management, and responsibilities are fragmented both in the horizontal and in the vertical management system. Thus it has been argued that the division of responsibilities between different actors makes the system rather diffuse and unclear (Østdahl, 2002). Despite this, the licensing procedures, as will be explained later, involve close coordination and interactions between laws and actual planning of water quantity and quality. It is however, important to notice that the integration of management systems should improve considerably with the implementation of the WFD. The WFD administrative system for water management will be based on hydrological basin boundaries and include different governmental sectors across the municipality and county boundaries in developing management plans aiming at holistic management of water resources.

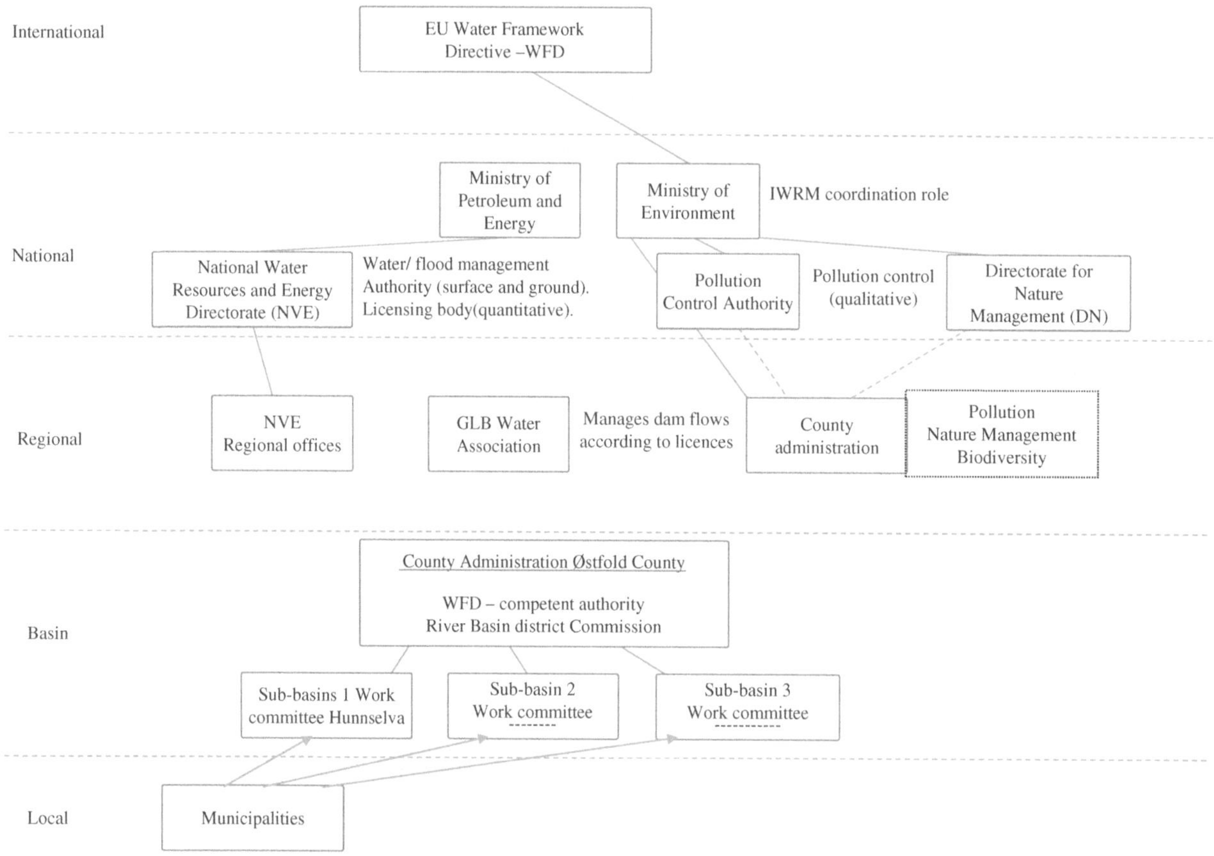


**Figure 2.3** Organogram that outlines the institutional structure in place for water management in the Glomma river basin. This is not intended to be exhaustive. Source: Campbell *et al.* (2009).

Despite a rather fragmented sectored management system at different spatial levels in Norway, the operation of the Glomma is fairly well managed and integrated. The local owners' association, GLB, has the operational responsibility for the total water resources, and regulates the water flow in order to maximize energy production, i.e., for the best possible economic results to their members, but within the framework of the licensed conditions. With regard to the implementation of the WFD, the GLB is represented on the Board of the Water District Authorities implementing the WFD and it is involved in the actual implementation of the Directive. The GLB (a WMA) acts as a link between power companies, the government and the public. GLB provides stakeholders with statistics and information related to the operation of river systems, maintenance of dam installations and follow-up of various license conditions.

#### *2.3.3.3 Public participation, access to information and decision making*

Making room for stakeholder interests to be expressed is nowadays common in the Norwegian institutional framework; but it has not always been a smooth process. Until about the 1950s, the water courses were perceived as unlimited resources and different sectoral interests could implement their action plans relatively independent of other interests. However, with accelerating hydropower development, rising levels of water pollution, and increasing recreational needs, environmental awareness increased. Conflicts between hydropower developers and environmental interests grew, starting in the 1960s and accelerating in the 1970s with confrontations such as those against regulation of the Mardøla river in 1970 and Alta in 1980 (Gleditsch, 1971; Hjorthol, 2006). These conflicts did not stop the projects themselves, but the protests eventually had major implications for the incorporation of conservation concerns in hydropower policy. Society became more concerned over environmental impacts, leading to a need for coordinated planning and allocation of water resources. Prior to this, applications had mainly been supported by technical plans. The first integrative initiatives included the Lake Mjøsa abatement program during the 1970s (see the earlier section on Lake Mjøsa), and the Master Plan for Water Resources in the 1980's. As mentioned earlier (Section 2.3.1), the Master Plan included a ranking system for the remaining undeveloped hydropower potential based on economic profitability and impacts for the different user interests. Stakeholders were widely involved in this exercise. The development of the master plans has provided guidelines, in parallel with the Plan and Building Act (Plan og bygningsloven, 1985), for an elaborate licensing process for water resources development. This licensing process has become a cornerstone in the decision-making process and a salient prerequisite is stakeholder involvement.

Traditionally, any initiative taken on hydropower construction (or other major works such as road/railway building) has to be presented by the executing organization to the general public, which can then submit their objections and/or propose modifications. When new hydropower projects are planned, public participation is normally financed by the municipality or in some cases by the counties. In the case of a new hydropower project, a notification (project description by the developer/applicant) is sent to the central and local authorities and also to the public for consultation. Local people may study the plans at the local post office, library or town hall. When and where the plans could be studied is advertised in the local newspaper. One or more public meetings are arranged in the project area to give information about the licensing process and the project plans. At the meetings, different opinions are expressed (including alternative plans) and the possible conflicts related to these plans or parts of the project are also discussed. Comments on special rights/use of water and area are of particular interest. Separate meetings to inform local administrations and politicians may also take place. Opinions concerning the application and the corresponding impact assessment analyses (IA) reports are sent to NVE (see Section 2.3.3.1 above). If the IA reports uncover new aspects, additional IA assessments may have to take place. A recommendation for a license will contain a set of conditions regulating the whole 'life' of a license: from approval of detailed construction plans regarding landscape to environmental and safety aspects, plant maintenance and even the ultimate, closing the plant. This could involve constructing weirs, building fish ladders, correcting river courses and removing vegetation from regulated zones. The conditions can be revised 30 to 50 years after the license has been granted. The owners of the power stations are also obliged by these conditions to pay annual fees to the municipalities and the state. In addition they may also be obliged to establish special funds to encourage local industry. Along with property and income taxes this makes some of these hydro power municipalities wealthy. Failure to adhere to the agreed conditions can result in a fine, and repeated violations may result in withdrawal of the license itself. According to the Norwegian legislation, affected land owners and farmers will be compensated either by mitigation measures or by money.

Besides this, the Ministry of the Environment and the Norwegian Climate and Pollution Agency (KLIF) also publish IWRM related data on the web. The new web-site for issues related to the WFD (www.vannportalen.no) also makes IWRM-relevant data available and the public can follow this work online for the Glomma Region. The Norwegian Directorate for Nature publishes "The State of Environment Norway", a web site and yearbook presenting data, indicators and goals for the environmental situation in Norway. In addition, publicly available databases are presently being issued. These include the database "Vannmiljø", which is envisaged to be the national database for water quality data; all water quality monitoring data must be registered in this database Also, the "Vann-nett" system offers maps of the water bodies in Norway, including information if the water bodies are at risk or not at risk of fulfilling the requirements of the WFD.

In terms of European legislation, Norway has ratified the UNECE Convention on Access to Information, Public Participation in Decision-making and Access to Justice in Environmental Matters (the Aarhus Convention) and is also implementing the Water Framework Directive, both of which provide for public participation in water management. The Freedom of Information Act (Offentlighetsloven, 2006) is the implementation of freedom of information legislation in Norway on a national level providing the rules and principles of access to public documents. The Environmental Information Act (Miljøinformasjonsloven, 2003), relates to the right to environmental information and public participation in decision-making processes relating to the environment. The purpose of this Act is to ensure public access to environmental information and thus make it easier for individuals to contribute to the protection of the environment, to protect themselves against injury to health and environmental damage, and to influence public and private decision-makers in environmental matters. Public participation in planning processes is regulated by Planning and Building Act (Plan og bygningsloven, 1985), the act which governs land use in general. Any initiative, be it related to hydropower, construction or road/railway building, has to be presented to the general public by the executing organization, who can then submit their objections/present modification. The implementation of the WFD will not lead to any major changes to this, as the Directive also requires the full participation of all users in the watershed.

### 2.3.4 Capacity building in the Glomma River basin

Capacity building in Norway is hardly ever mentioned in Norwegian laws and policy documents, but there are substantial capacity building initiatives in the Norwegian water administration and governance structure. The present on-going process of implementation of the WFD in Norway serves as an example of such initiatives. The implementation of the WFD has also resulted in a high interest from Norwegian water managers (mainly local and regional) to participate in various training courses and seminars organized by competent authorities, water-related associations and academic institutes (see e.g., the high interest in the national WFD conference organized in early 2010; http://www.vannportalen.no/hovedEnkel.aspx?m = 31139). The Ministry of Foreign Affairs (2008) recently proposed in a white paper that the capacities for water monitoring should be given increased resources in the range of 9–11 million Euro per annum, that 1.5 million Euro should be spent on research and development projects in addition to the existing resources. The government also highlighted the need for resources for citizen involvement, and that local participation in the planning processes should be strengthened. This should be achieved by increased support of 250–375 thousand Norwegian kronor per year to the existing map- and web-based information tool Vann-Nett (http://vann-nett.nve.no/innsyn/).

### 2.3.5 Transboundary issues and the Glomma River

The Glomma basin in Norway, unlike the three other case basins described in this book, is not a transboundary river system in the real sense. The basin is almost completely within the Norwegian borders, with only 1% in Sweden. The relationship between Sweden and Norway regarding the Glomma River basin has been characterized by very few problems and conflicts.

## 2.4 FINAL REMARKS

Historically, the Glomma River has been important for the local population with regard to nearly all aspects of their lives. However, hydropower development has been the most dominant interest in the basin, leading to a heavily regulated river. Water diversion for hydropower development has resulted in insufficient water flow in some places, resulting in a negative impact on the ecological conditions in the river, changes in biodiversity and a reduction of the fish stock. Most hydropower regulations in Glomma, however, now take into account some measures for environmental flow; although the applied flow levels are not necessarily satisfactory to all user interests. Conflicts occur between environmentalists and local people emphasizing recreational values (in particular fishing) on one side, and on the other side people concerned about economic development and employment opportunities in the region. Participation in decision making and transparency of information and management plans have mitigated conflicts and contributed to a more balanced water use. Knowledge building among stakeholders leads to better and more engaged participation and better decisions. These are two key elements in IWRM confirmed by STRIVER's work in Glomma. Examples of such transparent IWRM plans include among others, the Lake Mjøsa Action Plan (Mjøsaksjonen) and the Master Plan for Hydropower (Samlet Plan), and the recent (since 1980s) licensing procedures. The licensing process has become a cornerstone in the decision making process and a salient prerequisite is stakeholder involvement. Consequently, considerable emphasis has been placed on institutional capacity building, knowledge development, public awareness and participation and efforts to streamline the hydropower licensing process. In the Glomma River basin, capacity building is seldom mentioned in policy documents and legislation but there are nevertheless governance and institutional structures that have a high capacity for water management. Both the WFD and the licensing systems in Norway's national legislation have well developed frameworks for stakeholder participation. The current implementation of the EU Water Framework Directive is an important measure for the protection of the Norwegian river basins, and it is an important aspect in the process of implementing IWRM. The integration of management systems has already improved considerably with the WFD, as the different water related directorates and ministries now meet regularly to discuss the implementation strategies. Similarly, on a regional and local level, the need for more monitoring data, new river basin management plans and decisions on which measures will be most cost-effective makes it necessary that the management meet more often on water issues through the new fora that have been established (river basin committees, working groups, etc.). Also, the administrative system for water management is now based on hydrological basin boundaries aiming at a holistic management of water resources. It can be argued that the process of implementing IWRM in Norway started in the 1970s and 1980s with the Lake Mjøsa Action Plan and the Master Plan for Hydropower, as these were the first activities where stakeholders were integrated in the process, but it has taken a long time to fulfill requirements of the IWRM principles. As noted in this chapter however, there are reasons to believe that a full implementation of the WFD will advance the process of implementing IWRM in Norway.

## 2.5 REFERENCES


Berge, D., Nhung, D.K., Thu Hoand, P.T., Barton, D. and Nesheim, I. (2008) *The use of Environmental Flow in IWRM, with Reference to the Hydropower Regulated Glomma River in Norway and Sesan River in Vietnam/Cambodia.* STRIVER Report No. D 8.1

Campbell, D., Rieu-Clarke, A., Allan, A., Gooch, G.D., Stålnacke, P. and Nagothu U.S. (2009) *Final Stakeholder Analysis – Stakeholder Participation in the STRIVER Basins.* STRIVER Report D4.2. 81p. http://kvina.niva.no/striver/Disseminationofresults/STRIVERReports/tabid/80/Default.aspx

EC (2000) *DIRECTIVE 2000/60/EC OF THE EUROPEAN PARLIAMENT AND OF THE COUNCIL of 23 October 2000 establishing a framework for Community action in the field of water policy.*72p

Eie, J.A., Faugli, P.E. and Aabel, J. (1996) *Elver og vann Vern av norske vassdrag.* Noregs vassdrags- og energiverk. Grøndahl og Dreyers Forlag AS, Oslo.

Energiloven (1990) *LOV 1990-06-29 nr 50: Lov om produksjon, omforming, overføring, omsetning, fordeling og bruk av energi m.m. (energiloven).* http://www.lovdata.no/all/hl-19900629-050.html

Fazi, S. and Lo Porto, A. (eds.). Contributions by: Barkved, L.J., Lo Porto, A., Fazi, S., Raju, K.V., Latha, N., Manasi, S., Deelstra J., Thaulow, H., Paranjape, S., Joy, K.J. (2009) Scientific report on pollution source assessment, including source apportionment results, and pollution prevention measures STRIVER Report No. D 7.1.

Forurensingsloven (1981) *LOV 1981-03-13 nr 06: Lov om vern mot forurensninger og om avfall (Forurensningsloven).* http://www.lovdata.no/all/hl-19810313-006.html

Gleditsch, N.P., Hartmann, Å. and Naustdalsli, J. (1971) *Mardøla-aksjonen,* Institutt for fredsforskning, Oslo.

Hveding, V. (1992) *Hydropower development in Norway.* Institute of technology, Department of hydraulic engineering. NTNU, Trondheim.

Hjorthol, L.M. (2006) *Alta: kraftkampen som utfordret statens makt.* Gyldendal akademisk forlag. Oslo.

McNeill, D. (1998) Water as an economic good. *Natural Resources Forum* **22**(4), 253–261.

Miljøinformasjonsloven (2003) *LOV 2003-05-09 nr 31: Lov om rett til miljøinformasjon og deltakelse i offentlige beslutningsprosesser av betydning for miljøet (miljøinformasjonsloven).* http://www.lovdata.no/all/hl-20030509-031.html

Nesheim, I., McNeill, D., Stålnacke, P., Sekhar, N.U., Grizzetti, B., Allen, A. A., Barton, D., Begueria-Portugés, S., Berge, D., Bouraoui, F., Campbell, D., Deelstra, J., García-Ruiz, J.M., Gooch, G.D., Joy K.L., Lana-Renault, N., Lo Porto, A., Machado, M., Manasi, S., Nhung, D.K., Paranjape, S., Portela, M.M., Rieu-Clarke, A., Saravanan, V.S., Thaulow, H. and Vicente-Serrano, S. (2009) *The first IWRM assessment report for the four case basins: Glomma, Tagus, Sesan and Tungabhadra.* STRIVER Report No. D5.1

NVE (2004). *Guideline for preliminary determination of heavily modified water bodies in Norway* (in Norwegian). NVE, 31.05.2004, 20 pp. (http://www.nve.no/no/Vann-og-vassdrag/Vassdragsmiljo/Vanndirektivet/Sterkt-modifiserte-vannforekomster/)

Offenlighetsloven (2006) *LOV 2006-05-19 nr 16: Lov om rett til innsyn i dokument i offentleg verksemd (offentleglova).* http://www.lovdata.no/all/hl-20060519-016.html

Østdahl, T. Skurdal, J., Kaltenborn B.P. and Sandlund, O.T. (2002) Possibilities and constraints in the management of the Glomma and Lågen River basin in Norway. *Arch. Hydrobiol. Suppl.* 141/3–4, 471–490.

Plan- og bygningsloven (1985) *LOV 1985-06-14 nr 77: Plan- og bygningslov.* http://www.lovdata.no/all/hl-19850614-077.html

Qvenild, T. and Linløkken, A. (1989) *Glomma River – fish and hydropower.* Main report from the Glomma project.

Rogers, P., Bhatia, R. and Huber, A. (1998) *Water as a social and economic good: how to put the principle into practice.* Global Water Partnership technical advisory committee, Background paper no.2. Stockholm.

Saltveit, S.J. (2006). *Økologiske forhold i vassdrag – konsekvenser av vannføringsendringer. En sammenstilling av dagens kunnskap.* Norges vassdrags- og energidirektorat, Oslo.

Skarbøvik, E., Glover, B., Barton, D.N., Brabrand, Å., Bækken, T., Halleraker, J.H., Johansen, S.W., Kristiansen, A. and Saltveit, S.J. (2006) *Suggested methodology to establish environmental goals in Heavily Modified Water Bodies.* (In Norwegian with English summary)"Forslag til metodikk for fastsettelse av miljømål i sterkt modifiserte vannforekomster.". NIVA-Rapp. 5266. 84 s.

Stanhill, G. (1986) Water use efficiency. *Advances in Agronomy* **39**, 53–85.

The Ministry of Environment (1979) *The Mjøsa campaign.* The Ministry of Environment, Oslo.

The Ministry of Foreign Affairs (2008) *St.prp. nr. 75, (2007–2008), Om samtykke til godkjennelse av EØS-komiteens beslutning nr. 125 av 28. september 2007 om innlemmelse i EØS-avtalen av direktiv 2000/60/EF av 23. oktober 2000 om fastsettelse av rammer for Fellesskapets vannpolitikk (vanndirektivet)* http://www.regjeringen.no/nb/dep/ud/dok/regpubl/stprp/2007-2008/stprp-nr-75-2007-2008-.html?id = 519840&epslanguage = NO).

Vannresursloven (2000) *LOV 2000-11-24 nr 82: Lov om vassdrag og grunnvann (vannressursloven).* http://www.lovdata.no/all/hl-20001124-082.html

World Commission of Dam (2000) *WCD case study. The Glomma and Laagen River basins in Norway.* Final report 2000.

# Chapter 3

# Modeling water nutrient pollution with stakeholders involvement

B. Grizzetti, F. Bouraoui, L.J. Barkved and J. Deelstra

## 3.1 INTRODUCTION

Participatory processes are useful for improving multi-level environmental governance (Rauschmayer *et al.*, 2009). In the case of Integrated Water Resource Management (IWRM), participatory watershed modeling has been proposed as a means of addressing the complex issue of water management under contrasting interests. By including stakeholders' perspectives, participatory watershed modeling allows local knowledge to be considered and facilitates the acceptance of scientific modeling outcomes. This improves dialogue between science, policy and society and can contribute to the legitimacy of a decision making process (Korfmacher, 2001).

In the EC FP6 STRIVER project (www.striver.no), the potential for integrating stakeholders' views when modeling water nutrient pollution was explored. We analyzed water pollution issues in the Glomma in Norway and the Tungabhadra in India in the context of a twinning exercise, by applying the SWAT (Arnold *et al.*, 1998) model to both river basins. (The Tungabhadra modeling is given in Chapter 10 where also the general characteristics of the SWAT model are described). In the modeling process, we included the knowledge derived from the dialogue with local stakeholders during the annual stakeholder meetings organized by the project. In particular, we tried to articulate the interaction through three sequential phases of the modeling exercise: (1) modeling set-up, (2) modeling validation and (3) analysis of scenarios (Figure 3.1).

This approach has proven to be suitable for supporting IWRM when addressing the specific issue of water pollution by nutrients (Grizzetti *et al.*, 2009), and has provided particularly appropriate results in the Norwegian case in relation to the forthcoming implementation of the Water Framework Directive (WFD) (Barkved *et al.*, 2009). The physically-based SWAT model was applied in the Hunnselva and Lena catchments, which are part of the Glomma basin, to represent the actual pressures of nitrogen and phosphorus on water and predict the potential impact of future scenarios. The modeling study was developed by including inputs from local stakeholder groups participating in the STRIVER project. Scenarios were focused on the implementation of measures to reduce nutrient losses to water and on agricultural changes to produce more profitable crops.

This chapter presents the results of this research activity according to the three step approach as shown in Figure 3.1, with a particular emphasis on the biophysical modeling results, and discusses the relevance of stakeholders' involvement in the process. For general characteristics of the Glomma river basin, see Chapter 2.

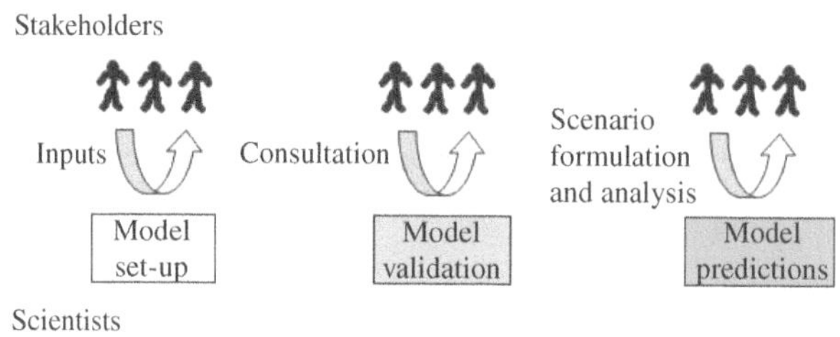


**Figure 3.1** Interactions between scientists and stakeholders in the modeling process.

## 3.2 DATA CONSTRUCTION AND MODEL SET-UP

### 3.2.1 Data availability

The application of the SWAT model (Arnold *et al.*, 1998) requires the collection of several types of data on the study catchment, including geographically referenced data, time series of measured runoff, nutrient concentrations, climatological data, and farm management information. In modeling, the time frame and the spatial resolution are set according to the objectives of the study but are often restricted by the resolution of available data. This trade off has highly influenced the model application in the Hunnselva and Lena catchments. In fact, although the objective was the evaluation of present and future nutrient pressures, the available data concerning runoff, nutrient concentrations, land use and farming practices still did not cover all the data requirements for optimal use of the model, a situation likely to be encountered by river basin managers. A specific advantage of modeling in this case is the possibility of including hypotheses or expert estimations to compensate for the missing information.

For this reason, interactions with local stakeholders, promoted during the STRIVER project through field visits and annual meetings, was particularly beneficial in overcoming some data limitations, providing expert estimates to fill in the lacking data.

### 3.2.2 Modeling set-up

Considering the time period covered by the available data on weather, water flow, water quality, it was decided to run the SWAT model from 01/01/1997 to 31/12/2006. The first year was kept for model initialization, the period 1998–2001 was considered for the calibration, and the period 2002–2006 for the validation. Daily water flow was calibrated at the station Lena 2.634.1000.1, which is situated in the middle of the Lena catchment, since no other water flow measures were available in the area of study during the period 1997–2006 (Figure 3.2). The average annual water flow produced by the Hunnselva and Lena catchments was estimated using measurements prior to 1997 (from Hunnselva 2.619.0.1000.1 and Skreia_2.466.0.1001.0 gauging stations). Bimonthly measurements of total nitrogen and total phosphorus were available for the period of study at the outlet of the two catchments (Figure 3.2). Nitrogen and phosphorus were calibrated considering annual and average monthly values. The sediments were not calibrated, since measurements were not available at any station. This imposed a strong limitation in the calibration of nutrients, in particular for phosphorus, which is transported with sediments as particle bound phosphorus.

#### *3.2.2.1 Spatial discretisation*

The river network and the sub-basins delineation was based on the digital elevation model (DEM) from the Norwegian Mapping Authority with resolution 25 × 25m. The threshold area for stream definition in SWAT was set to 200 ha, to account for the high river density of the region. A total of 17 sub-basins were defined, 6 in the Hunnselva catchment and 11 in the Lena catchment (Figure 3.3). One managed reservoir was modeled in the Hunnselva catchment (Figure 3.3). Scattered dwelling and point sources were entered for each sub-basin. Within the sub-basins, the Hydrological Response Units (HRU) were established considering unique combinations of land cover, soil and slope classes with surface contributions higher than 1%, 10% and 30%, respectively. This resulted in the definition of 345 HRU for the entire region of study.

#### *3.2.2.2 Land cover and crop types*

A land cover map from the Norwegian Forest and Landscape Institute was included in the modeling. According to the land cover map, agricultural land occupied 25% of the surface of the Hunnselva-Lena catchment. The distribution of crop types within the agricultural land was established based on the statistics available from the Norwegian Forest and Landscape Institute, data provided by Norwegian Statistics and the Food and Agricultural Organization's (FAO) crop statistics for Norway for the year 2000.

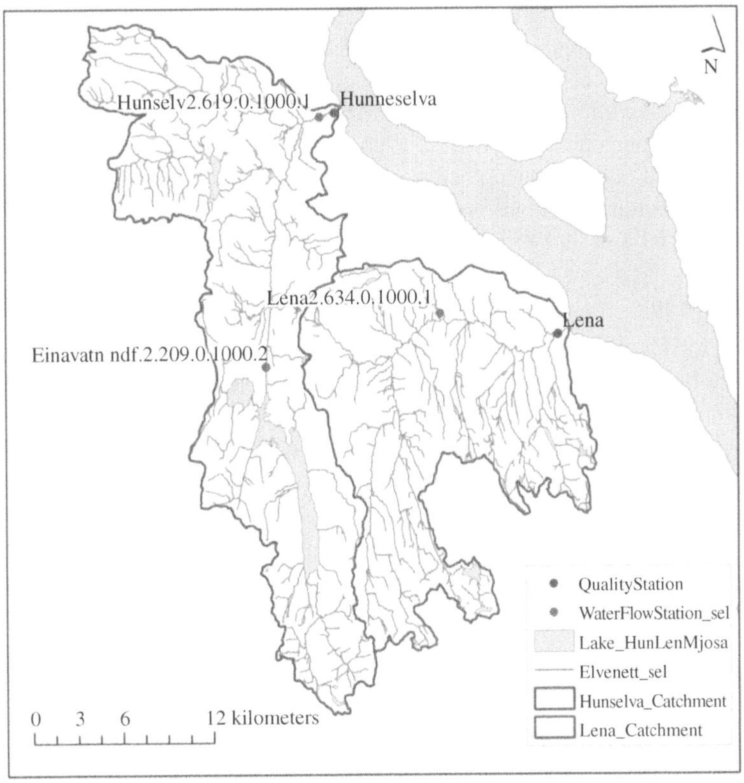


**Figure 3.2** Water flow and quality gauging stations in the Hunnselva and Lena catchments.

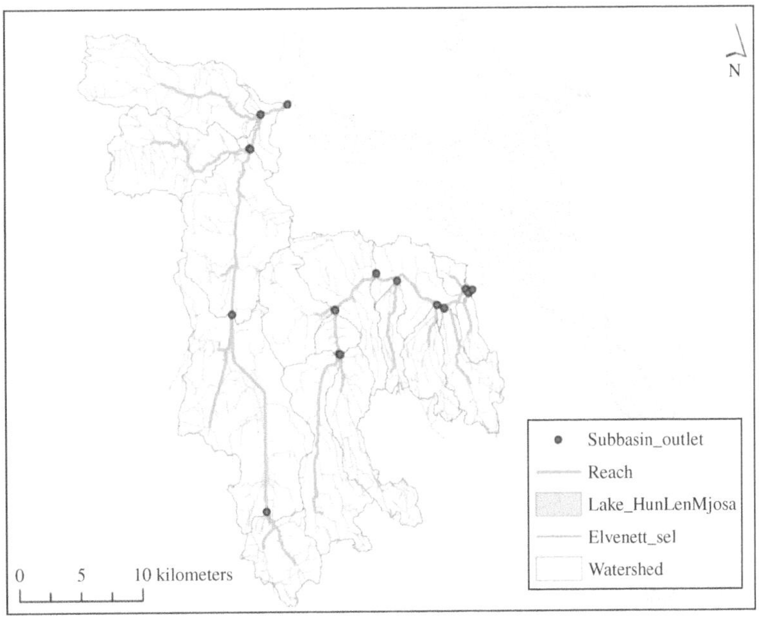


**Figure 3.3** River network and sub-basins delineation in the SWAT model for the Hunnselva and Lena catchments.

### *3.2.2.3 Soils*

For the total Hunnselva and Lena watersheds no digitized soil map suitable for SWAT modeling was available. It was therefore necessary to develop a new map using the information available from two sources:

(1) the map of agricultural soils (provided by the Norwegian Forest and Landscape Institute), which includes a detailed description of soil characteristics per agricultural parcel; and
(2) the FAO Digital Soil Map of the World.

The first was considered for the soils in the agricultural land, while the latter was used to describe the remaining soils, mainly covered by forest. In the modeling, a general soil depth of 100 cm was considered, 30 cm for the first layer and 70 cm for the second. The soil physical parameters of the dominant soil in the non-agricultural area were calculated according to the FAO Soil Map, while for the agricultural areas these were calculated using pedotransfer functions specifically developed for Norwegian conditions (Riley, 1996). Additional parameters required by SWAT for computing soil erosion were calculated based on relationship between texture and soil properties available in literature (Maidment, 1993). The majority of the agricultural soils are provided with an artificial subsurface drainage system, having a drain spacing between 8 and 10 m, and a drain depth below the soil surface between 0.8 and 1.0 m.

### 3.2.2.4 Weather information

The weather data were provided by Bioforsk's climatic station at Kise, located outside the study area, close to the outlet of the Hunnselva catchment but on the opposite side of Lake Mjøsa (Figure 3.2). The data included daily precipitation, humidity, global radiation, wind speed, and maximum and minimum temperature (Figure 3.4). Data used for the precipitation were obtained from Bioforsk's climatological station Apelsvol, located within the Lena catchment, as these measurements are more representative for the rainfall pattern in the Hunnselva and Lena catchment.

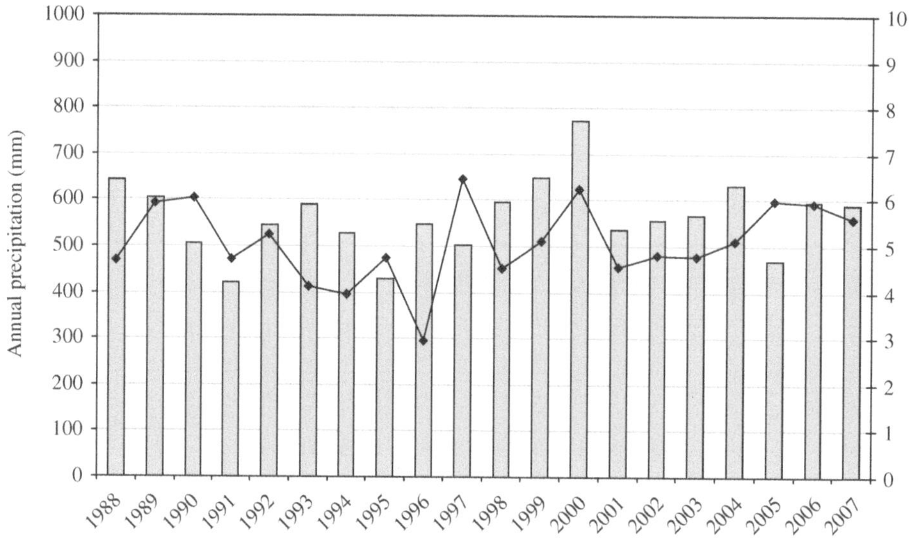


**Figure 3.4** Annual precipitation and average temperature at the weather station Kise.

### 3.2.2.5 Point sources and scattered dwellings

Nitrogen and phosphorus input from scattered dwellings were estimated from the population living in the three main municipalities covering the area of study: Ostre Toten, Gjovik and Vestre Toten. These sources of nutrients were entered as point sources for the respective sub-basins, and scattered dwellings were considered to be halved before entering into the river network at the end of each sub-basin.

### 3.2.2.6 Fertilizer applications and management operations

The model input concerning the fertilizer application and management operation for the different crop types was based on information provided by local stakeholders and completed by the Norwegian experts of Bioforsk and the Norwegian Institute for Water (NIVA); and the crop data available from the FAO database (for year 2000).

The Hunnselva catchment is characterized by livestock production, but little detailed information was available concerning animal types, numbers and the husbandry typology. However dairy cattle dominate in Norway and therefore milk cattle, kept indoors, were assumed to be the dominating form of animal husbandry. The manure produced was considered to be proportional to the crop needs and spread on barley and fodder, according to the annual application rates provided by the experts for these crops. These assumptions may however lead to overestimations of manure application, as livestock production is mainly concentrated in particular areas of the catchment.

The information available on fertilizer application for modeling purposes is reported in Table 3.1 and Table 3.2, for mineral fertilizer and manure, respectively. The values consist of averages for the period 1991–2005 and were set according to the data available for the Kolstad nearby catchment, in addition to recommended fertilizer applications relative to yield expectation provided by Bioforsk/Apelsvol (Table 3.1).

**Table 3.1** Mineral fertilizer application and expected yield per crop.

| Crop | N (kg/ha) | P (kg/ha) | K (kg/ha) | Expected yield (kg/ha) |
|---|---|---|---|---|
| Barley | 95 | 14 | 70 | 4000 |
| Rapeseed | 102 | 45 | | |
| Grass/fodder | 150 | 16 | 85 | |
| Broccoli | 180 | 40 | 160 | 8000 |
| Potato | 100 | 45 | 155 | 30000 |
| Strawberry | 60 | 15 | 50 | |

**Table 3.2** Manure application per crop considering dairy fresh manure.

| Crop | Manure (kg/ha) | N (kg/ha) | P (kg/ha) |
|---|---|---|---|
| Barley | 1000 | 38 | 8 |
| Grass/fodder | 1316 | 50 | 11 |

According to the mineral and manure fertilizer rate per crop type and the area occupied by each crop, the amount of nitrogen and phosphorus input in agriculture was estimated for the Hunnselva and the Lena catchment, together with the calendar of management operations established for the modeling (baseline scenario).

## 3.3 MODEL CALIBRATION AND VALIDATION

### 3.3.1 Water flow

The daily water flow was calibrated for the period 01/01/1998-31/12/2001 and validated for the period 01/01/2002-31/12/2006 at the station Lena 2.634.1000.1 (Figure 3.2). The Nash-Sutcliffe efficiency (Nash and Sutcliffe, 1970) for the daily water flow simulation was 0.40 for the calibration and 0.60 for validation (Figure 3.5). In general, the total water amount and the flow peaks are well simulated, except for the flood event at the end of year 2000.

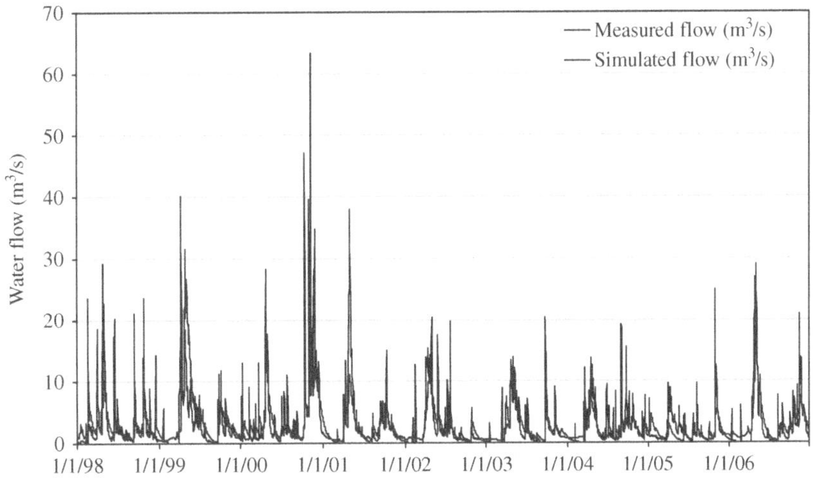


**Figure 3.5** Measured and simulated daily water flow at the station Lena 2.634.1000.1.

The simulated long term average discharge at the outlet of the Hunnselva and Lena catchments were 5.5 $m^3$/s and 4.4 $m^3$/s, respectively. These values are in good agreement with the available average discharge of 5.5 $m^3$/s (average of the period 1986–1993 at the station Hunnselva 2.619.0.10001.0) and 3.9 $m^3$/s (average of the period 1986–1995 at the station Skreia 2.466.0.1001.0), although these discharges refer to a different time period.

### 3.3.2 Water quality

Total nitrogen and phosphorus concentration measurements were available from the Hunnselva and Lena catchment outlets. However, the calibration of water quality was problematic because no information was available on sediments and on water flow at the catchments outlet.

For water quality calibration, annual and monthly values were considered. An analysis of the model monthly simulations revealed an over prediction of nitrogen and especially phosphorus associated with high predicted sediments peaks. The parameters associated with sediment production were adjusted accordingly to reduce the erosion and increase sedimentation in the reservoir. Figure 3.6 shows the results for annual nitrogen and phosphorus predictions at the Hunnselva and Lena outlets. For nitrogen both concentration values and trends are well represented, while for phosphorus the model tends to overestimate the concentration. A possible explanation could be the amount of manure considered in the modeling or the overestimation of the sediment transport. In fact, most of the phosphorus is transported attached to soil particles, while nitrogen is much more soluble in water. The amount of manure considered in the modeling depends on the assumption that all barley and grassland are fertilized with manure. However, probably this does not correspond to the reality in all the catchment, since livestock are concentrated in some areas. The overestimation of manure application may lead to a higher fraction of organic nitrogen and especially phosphorus simulated by the model.

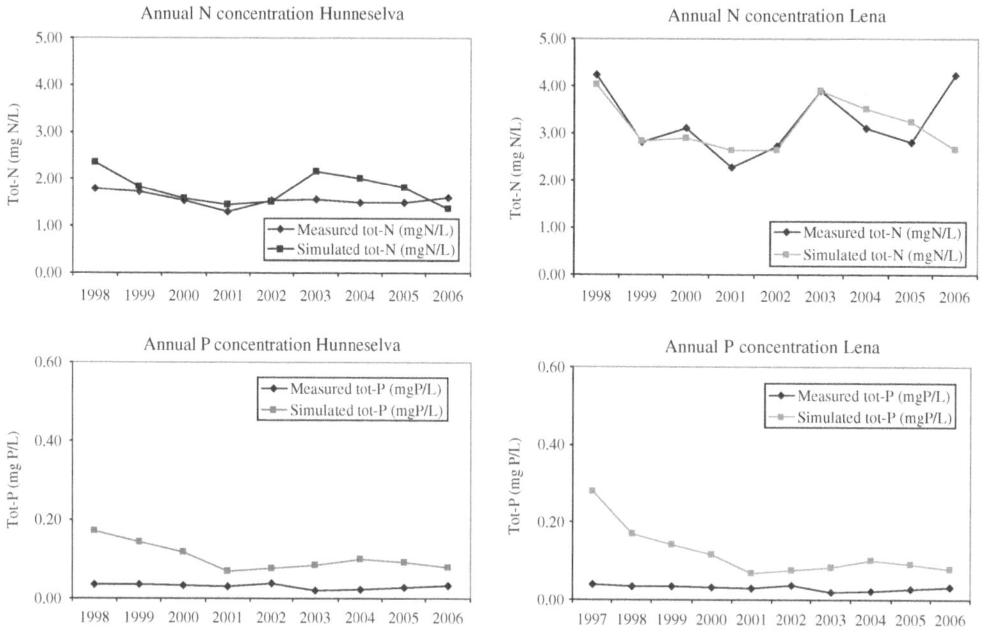


**Figure 3.6** Measured and simulated annual total nitrogen and phosphorus concentrations at the Hunnselva (left) and Lena (right) outlets.

Furthermore, uncertainty exists whether the concentration based on a fortnightly grab sampling routine can represent real concentrations of nitrogen and phosphorus. Phosphorus loss is very much linked to erosion, and is to a much larger degree than nitrogen, concentrated in periods with high runoff. Runoff, especially in agricultural dominated catchments, is characterized by large diurnal variations. These periods are also often characterized by large values of phosphorus and suspended solid concentrations which are not, or are poorly represented, when sampling is based on a fortnightly sampling. Monthly average values based on grab samples might well under-represent the real values for nitrogen and phosphorus concentration. However in our case this information was lacking. It is also important to consider that most erosion models used in Europe are designed to estimate erosion caused by rainfall induced runoff (Van-Camp *et al.*, 2004). Such models cannot predict erosion from snowmelt and, consequently Lundekvam (2002) has developed an erosion model – ERONOR – for Norwegian conditions, and for Finland the ICECREAMS-model is being used. The development of these models illustrates the need for process-based models appropriate for the dominant erosion process. Therefore, in the application for scenario analysis, the SWAT model predictions should be considered more as an indication of trends than of absolute values.

## 3.4 ANALYSIS OF SCENARIOS

### 3.4.1 Development of scenarios

The scenarios were developed jointly by scientists and the local stakeholder group participating in the STRIVER project. The scenarios focused on the implementation of measures to reduce nutrient losses to water and on the effect of agricultural changes towards more profitable crops. Three types of measures related to the type and timing of ploughing, the amount of fertilizer application and the changes in crop type proportions were tested. In particular, the scenarios included:

(1) Scenario 1 – Reduced ploughing – for grain crops, in particular barley, rapeseed and fodder, the autumn ploughing is substituted by a light harrowing.
(2) Scenario 2 – Optimal fertilization – the crop fertilizer applications (Table 3.8 and Table 3.9) are substituted by optimal fertilizer application estimated by the model, according to the actual crop needs.
(3) Scenario 3 – Increase in vegetables/berry production - the growing of more profitable crops is extended. Barley is reduced from 15% (baseline) to 10% of the catchment area, while vegetables/berry (including broccoli, potato and strawberry) are extended from 3% (baseline) to 8%.

## 3.4.2 Results of scenarios

The annual nitrogen and phosphorus concentrations and sediment production predicted by the model at the catchment outlet were considered as indicators to evaluate the impact on water quality of specific scenarios. The results of the scenarios application are presented in Table 3.3 and Table 3.4 for Hunnselva and Lena catchments, respectively.

**Table 3.3** Results of scenarios implementation in the Hunnselva catchment, predicted by the SWAT model.

| Average for 1998–2006 | Concentration tot-N (mgN/L) | Concentration tot-P (mgP/L) | Sediment production (ton/ha) |
|---|---|---|---|
| **Baseline** | 1.79 | 0.10 | 0.15 |
| **Scenario 1** Reduced ploughing | 1.61 | 0.08 | 0.12 |
| **Scenario 2** Optimal fertilization | 1.40 | 0.10 | 0.15 |
| **Scenario 3** Increase in vegetables/berry | 2.00 | 0.10 | 0.15 |

**Table 3.4** Results of scenarios implementation in the Lena catchment, predicted by the SWAT model.

| Average for 1998–2006 | Concentration tot-N (mgN/L) | Concentration tot-P (mgP/L) | Sediment production (ton/ha) |
|---|---|---|---|
| **Baseline** | 3.14 | 0.17 | 0.37 |
| **Scenario 1** Reduced ploughing | 2.74 | 0.12 | 0.25 |
| **Scenario 2** Optimal fertilization | 2.01 | 0.16 | 0.37 |
| **Scenario 3** Increase in vegetables/berry | 3.30 | 0.17 | 0.32 |

### *3.4.2.1 Scenario 1 – Reduced ploughing*

The first scenario simulates the application of more conservative tillage, to reduce soil erosion. This type of scenario is in line with the implementation of good management practices to conserve the soils, as recommended by the EC rural development policy. For the grain crops (barley, rapeseed and fodder) the autumn ploughing is substituted by a light harrowing. The results of model prediction show that the erosion is reduced as expected and that total nitrogen and phosphorus concentrations decrease, due to the decline of the organic fraction, mainly transported with sediments (Table 3.3 and Table 3.4). For phosphorus, the effect of conservation tillage is as expected in Norway, this also being the main reason for the large scale implementation of this tillage method in Norway. However, the modeling results for nitrogen need further investigation as the nitrogen mainly is lost as nitrate through the subsurface drainage system and is to a large degree not affected by conservative tillage.

### *3.4.2.2 Scenario 2 – Optimal fertilization*

The second scenario simulates the optimization of fertilizer application according to the actual crop needs, in order to balance and reduce the use of fertilizer. Such an optimization of fertilizer application is foreseen by the EC Nitrates Directive, especially in Nitrate Vulnerable Zones. In the modeling only the mineral fertilizer was optimized, while the manure was kept constant, under the hypothesis that farmers have to spread out in the fields manure produced by their livestock. The second scenario results in a decrease in the nitrogen loss, but does not affect significantly the phosphorus (Table 3.3 and Table 3.4). This could be due to the fact that phosphorus, contrary to nitrogen, is attached to soil particles.

#### *3.4.2.3 Scenario 3 – Increase in vegetables/berry production*

The third scenario was established to evaluate the impact on water quality of land use changes driven by the local market. According to this scenario, the growing of more profitable crops is extended. Barley is reduced from 15% (baseline) to 10% of the catchment area, while vegetables/berry (including broccoli, potato and strawberry) is extended from 3% (baseline) to 8%. The fertilizer application rate per crop type was maintained the same as in the baseline. For the Hunnselva, the results of model prediction show that nitrogen loss increases due to the higher fertilization rate required by broccoli compared to barley. This is not immediately evident for phosphorus. However, in Lena, it appears that sediment production decrease as a consequence of the land cover change, without affecting the phosphorus concentration. These results contradict the general consensus in Norway that vegetable production increases both the erosion and P-loss, especially when considering the significant increase in P application the vegetable crops compared to barley. However two aspects should receive further consideration. First, the effects of land use variation on sediments and nutrients concentration are much more evident if the changes occur closer to the measurement point (the catchment outlet in this case), which was the case in Lena. Secondly, the presence of a big reservoir in Hunnselva acts as a buffer for sediments and nutrients transported with sediments, reducing the response in downstream water concentration.

## 3.5 ADVANTAGES OF INVOLVING STAKEHOLDERS

Participatory modeling appears to be particularly appropriate for IWRM, due to the potential for combining technical knowledge with economic and social concerns related to water resource use and protection (Voinov and Gaddis, 2009). The strengths of participatory modeling lie in the flexibility of the approach and its potential for structuring the active involvement of stakeholders, offering an opportunity for shared language and knowledge integration (Veidera *et al.*, 2009).

In STRIVER we worked together with local stakeholders of the Hunnselva and Lena catchments during different phases of the modeling activity, establishing a co-operation with the WFD working group. During the project period, interactions developed successfully. This was mainly due to several encounters between local managers and scientists, which allowed for the development of a gradual mutual understanding of the scientific and management aspects of the research. Such an approach allowed topics to be addressed and discussed several times, ensuring interest, relevance and reciprocal understanding amongst the group. During the stakeholder workshops there was a strong willingness and interest among the local stakeholders to participate in and contribute to STRIVER activities.

The participation of stakeholders was found to be beneficial during all of the different phases of the watershed modeling activity. At the beginning of the project their contribution was crucial to overcome the problem of a significant lack of data for the area of study, especially related to farming practices, such as amount of fertilizer applications, type of crops and management practices, providing educated guesses or expert approximations of information otherwise not available for the study. Their involvement in the discussion raised awareness among stakeholders of the importance of data collection and of coherence of monitoring plans. Finally, these interactions helped the modelers to focus the study on pollution issues relevant to the area.

During the second and third workshops the dialogue between STRIVER researchers, local managers and end users continued and improved. In addition, jointly with stakeholders and local managers, field visits were carried out to obtain an on-sight impression of several watershed management issues. The results from model simulation were presented and potential management scenarios to be implemented in agriculture were discussed. This interaction proved to be very fruitful for the parties involved. Stakeholders understood the potential of modeling tools and helped formulate scenarios of relevant and realistic alternative management strategies, as they were interested both in the effects of the incoming legislation on nutrient pollution and of the market driven land use changes. Similarly, the dialogue was very useful also for scientists, as they received feedback on the reliability of their modelling results and were further guided on performing scenarios analysis and evaluating the outcomes. Certainly, maintaining active participation and assuring correct representation of interests are underpinning issues in participatory modeling (Veidera *et al.*, 2009), but these problems were limited in our case by the local scale of the study.

Ultimately, the stakeholder workshops helped in fostering linkages and trust between the STRIVER researchers, managers, end users and policymakers, and at the same time improved transparency and acceptance of the project outcomes.

## 3.6 CONCLUSIONS

The application of the SWAT model to the agricultural Norwegian catchments of Hunnselva and Lena, in the Glomma river basin, has shown promising results concerning the implementation of SWAT as a potential tool for assessing the nitrogen and phosphorus pressures on waters and to promote IWRM through the involvement of stakeholders in the modeling process. Limitations of the model application have emerged but these might have been caused due to the dominance of winter conditions in the generation of runoff, nutrient and soil loss. Also in a relative data rich country like Norway, data was lacking for the parameterization and calibration/validation of the SWAT model. This is an important lesson and demands an on-going communication between managers and scientist concerning data collection. However, under these conditions, the proposed approach has provided informative estimates, allowing identification of the major sources of pollution, their spatial distribution and the underpinning processes. The involvement of stakeholders in the modeling exercise improved the overall quality and relevance of the study, especially concerning the detailed collection of information on farming practices. Such information, retrieved via a continuous dialogue with the local farmers, was crucial to assess water pollution. Similarly, the interaction with stakeholders contributed to the development of scenarios relevant for the local water pollution issues and economic interests. For these reasons, the modeling approach with stakeholders' involvement proposed in the STRIVER project answered specific water manager and stakeholder interests related to water pollution resulting in an effective approach to WFD implementation and promotion of IWRM.

## 3.7 REFERENCES


Arnold, J.G., Srinivasan, R., Muttiah, R.S. and Williams, J.R. (1998) Large area hydrologic modeling and assessment part I: model development. *Journal of the American Water Resources Association* **34**, 73–89.

Barkved, L.J., Deelstra, J., Grizzetti, B. and Bouraoui, F. (2008) Modelling nutrients in the Glomma river basin: scenarios & management recommendations. STRIVER Policy Brief No. 11, http://kvina.niva.no/striver/Portals/0/documents/STRIVER_PB11.pdf

Grizzetti, B., Lo Porto, A., Barkved, L.J., Bouraoui, F., Deelstra, J. and Joy K.J. (2008) Modelling water pollution with stakeholders involvement – The twinned experience of Glomma (Norway) and Tungabhadra (India) river basins. STRIVER Policy Brief No. 10, http://kvina.niva.no/striver/Portals/0/documents/STRIVER_PB10.pdf

Korfmacher, K.S. (2001) The politics of participation in watershed modelling. *Environmental Management* **27**(2), 161–176.

Lundekvam, H., 2002. ERONOR/USLENO – Empirical erosion models for Norwegian conditions. Documentation. Report no 6/2002. Norwegian University of Life Sciences, http://www.umb.no/eacc/article/model-documentation-download

Maidment, D.R. (ed.) (1993) *Handbook of Hydrology*. McGraw-Hill, Maidenhead.

Nash, J.E. and Sutcliffe, J.V. (1970) River flow forecasting through conceptual models *Journal of Hydrology* **10**, 282–290.

Rauschmayer, F., Paavola, J. and Wittmer, H. (2009) European governance of natural resources and participation in a multi-level context: an editorial. *Environmental Policy and Governance* **19**, 141–147.

Riley, H. (1996) Estimation of soil physical properties of cultivated soils in southeast Norway from readily available soil information. *Norwegian Journal of Agricultural Sciences* **25**(Supp), 51

Van-Camp,. L., Bujarrabal, B., Gentile, A.-R., Jones, R.J.A., Montanarella, L., Olazabal, C. and Selvaradjou, S-K. (2004) Reports of the Technical Working Groups Established under the Thematic Strategy for Soil Protection. EUR 21319 EN/2, 872 pp. Office for Official Publications of the European Communities, Luxembourg.

Videira, N., Antunes, P. and Santos, R. (2009) Scoping river basin management issues with participatory modelling: The Baixo Guadiana experience. *Ecological Economics* **68**, 965–978.

Voinov, A. and Gaddis, E.J.B. (2008) Lessons for successful participatory watershed modelling: a perspective from modelling practitioners. *Ecological Modelling* **216**, 197–207.

# Chapter 4

# Pressure-impact multi-criteria environmental flow analysis: Application in the Øyeren delta, Glomma River basin, Norway

*D.N. Barton, D. Berge and R. Janssen*

## 4.1 INTRODUCTION

Multiple criteria analysis improves the documentation of expert judgment in the establishment of environmental flow criteria. Trade-offs between river flow regimes, water levels and user interests can be explored using an adaptation of value functions which we have called 'pressure-impact curves for environmental flow' (PIMCEFA), based on a pooling of multiple expert judgments. They differ from value functions often seen in Multi-criteria Analysis (MCA) in that environmental flow is a common driver across all impacts. Pressure-impact curves are expert judgments of biophysical and socio-economic impacts of different environmental flow levels. Because environmental flow is already on a common scale ($m^3/s$, or water level), expert judgment can be limited to 'scientific' evaluation of impact, avoiding subjective valuation judgments of the importance of a particular impact relative to another. All subjective assessment is confined to a subsequent 'weighting step' of MCA. Weighting procedures in multi-criteria analysis offer a number of possibilities to explore the consequences of giving priority to specific experts and/or water user interests. Graphical visualization of impact curves also opens up the possibility of including 'layman' or stakeholder judgments in the analysis. Quantitative assessment of environmental impact and value judgments of the impacts' relative importance is normally blurred in MCA with the traditional approach to 'value functions'. The MCA software DEFINITE is used to demonstrate this separation of 'scientific' impact assessment and 'political' value judgment. The chapter also illustrates how pressure-impact curves can be used to quantify uncertainty presented by different experts. The argument is that assessment of low flow requirements in Norway has lacked a replicable approach to evaluating, and then comparing, alternatives across multiple social and ecological interests. We see multiple criteria analysis of environmental flows as an improvement at little extra cost of the 'traditional' approach to setting flow regulations in hydropower concessions, which in many countries has been based on largely undocumented 'stand alone' judgments by individual experts.

PIMCEFA is an approach to identifying impacts of different water levels and stream-flows, using so-called 'value functions', and a multi-criteria approach that allows decision-makers to evaluate trade-offs between different user and ecological interests. With a focus on social as well as ecological values, we define 'environmental flows' as water left in a river system, or released into it, for the purpose of maintaining certain river ecosystem services. For ease of presentation this chapter will use 'ecosystem service' to represent both 'ecological value' and 'user value'. Although all ecosystem services are conditional on ecosystem function, in multi-criteria analysis, river functions with no user interest *per se* provide 'regulating and supporting services (MEA, 2005). Such services may have 'no-use value', and be

©2010 IWA Publishing. *Integrating Water Resources Management: Interdisciplinary Methodologies and Strategies in Practice*. Edited by Geoffrey D. Gooch, Alistair Rieu-Clarke and Per Stålnacke. ISBN: 9781843393252. Published by IWA Publishing, London, UK.

presented as 'value functions' of a proxy indicator can be identified by experts. Making compensating trade-offs between different ecosystem services technically possible is both a strength and a weakness of MCA which we come back to in the final discussion.

In this chapter we evaluate how well a commercially available generic MCA software – DEFINITE (Janssen and Herwijnen, 2007) – can be adapted to assess specific issues of environmental flow. We discuss the separation of the roles of expert judgment and decision-maker weighting of interests. We argue that PIMCEFA responds to needs of a first level impact assessment by employing expert judgment (IUCN, 2003) combined with lessons from the established multiple criteria analysis literature, resulting in:

- better documentation of expert judgment of environmental impacts of a range of different hydrological regimes;
- a quantitative approach which can work even in data scarce settings which lends itself to replication by third parties;
- a generic method that is flexible enough to be adapted to many water management decisions, and lends itself to periodic revision at low cost as new information becomes available.

The focus here is on a technical explanation for and evaluation of PIMCEFA. For a discussion of the role of multi-criteria environmental flow evaluation in interactions between science, policymakers, and stakeholders, see Berge *et al.* (2010). We first provide a stepwise overview of PIMCEFA; we then discuss each step using illustrations of the data from the DEFINITE software screen shots; we then discuss the pros, cons and potential improvements of PIMCEFA based on a 'self-evaluation' of the experts who tried the method; with this overview we compare PIMCEFA with other multi-criteria approaches for determining stream flow; finally, we present some conclusions regarding the relevance of PIMCEFA for hydropower management in Norway.

## 4.2 TRADITIONAL APPROACH TO EXPERT ASSESSMENT OF ENVIRONMENTAL OR COMPENSATION FLOW IN NORWAY

More than 200 methods to assess Environmental Flows in regulated rivers have been described in the literature over the last years, grouping the methods into 4 or 5 types (Halleraker and Harby, 2006; Jowet, 1997; Dunbar *et al.*, 1998; Tharme, 2004). The EF methods are a mixture of subjective expert judgment and objective mathematical model calculations:

(1) Hydrological methods – EF is set as a percentage of natural flow.
(2) Hydraulic methods – These methods also include river morphology and water levels, but still no direct linkage to biology.
(3) Habitat simulation methods – Functional connections between physical alterations and river biology
(4) Holistic methods – Includes the flow needs of both the river ecosystem components, as well as the flow needs of the water use interests.

Berge *et al.* (2010) reviews the traditional approach to expert assessment of environmental flow in Norway. The formal process for river regulation in Norway depends on the size of the project. If the hydropower project is 40 GWh or greater, the project needs approval under the Plan and Building Law (PBL). This includes an Environmental and Social Impact Assessment, public and stakeholder involvement, hearings, etc. If the project is between 30 and 40 GWh, the Norwegian Water and Energy Directorate decides if approval in accordance with the PBL is necessary. If the project is below 30 GWh, the project does not come under the PBL approval, and approval is much simpler. For these smaller regulations there is a general requirement under the Water Resources Law (WRL) for a minimum flow of at least "common low flow" (not far from 10% of average flow). In the large projects, the question of minimum flow is decided in the terms of the hydropower concession or operating license – no specific requirements for the size of the flow are given by the WRL.

Most hydropower regulations in Glomma River basin (if we omit the pure Run of the River Regulations and the oldest regulations), have some kind of minimum release, compensation flow, or environmentally motivated rules for water level variations in reservoirs. The approach can be roughly described as an 'expert judgment based on baseline studies of ecological and water use interests', often simply referred to as the 'Expert Panel Method' in Norway. In the old regulations there was often only one expert that assessed all environmental aspects, frequently the regional Fish Inspector from the Directorate for Nature Management. In more recent regulations several experts and river users are involved in assessing the minimum water flows and the water levels. This expert group is often called an Expert Panel. The panel is composed of different experts on a case-by-case basis. The assessment methodology underlying expert judgment is often not documented in concession documents. IF this is the case when a concession comes up for renewal, revisions of compensation flow cannot be made with a replicable methodology. Multi-criteria environmental flow assessments discussed in this chapter are most relevant for larger projects. A full impact assessment mobilizes a number of different experts and a consistent approach to documenting their expert judgment. Furthermore, compensation flow aims at balancing environmental and social interests – multi-criteria methods offer a transparent and replicable way of weighting the importance of different interests in the determination of compensation/environmental flows.

## 4.3 OVERVIEW OF PIMCEFA

The main steps of PIMCEFA are shown in Figure 4.1. The approach broadly follows steps common to multiple criteria analysis (Beinat, 1997). However, some emphasis is placed on our interpretation of the roles of experts and stakeholders, criteria and ranking results as specific for the decision-making regarding environmental flow.

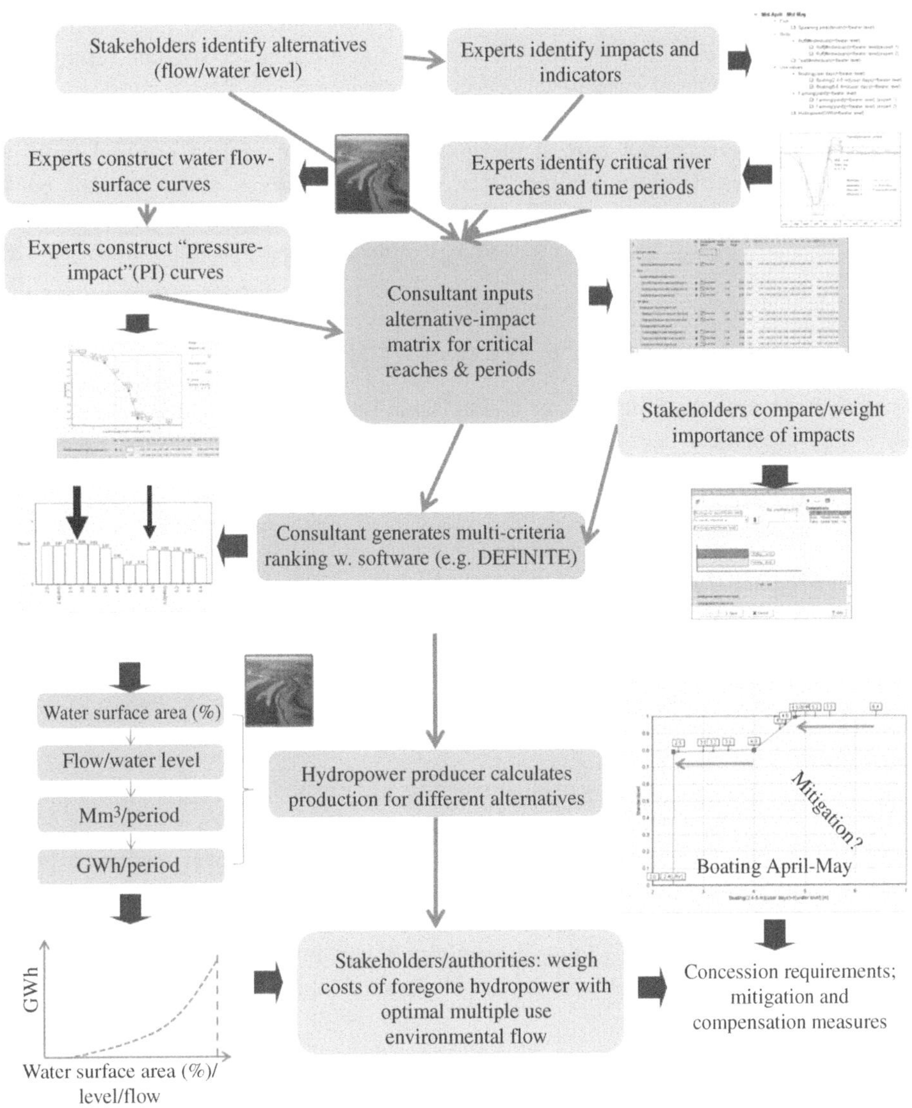


**Figure 4.1** Overview of the multi-criteria environmental flow method. Source: own elaboration Follow-up steps to this analysis include converting optimal multi-criteria water levels to flow curves and different hydropower production levels and their economic value. This final step (not illustrated in figure 1) allows one to evaluate opportunity costs to hydropower of different water levels proposed by the MCEFA (which would consider all impacts that could not be valued in economic terms). Source: own elaboration.

(1) **Stakeholders identify alternatives.** Identification of policy-relevant alternatives, ideally conducted by stakeholders. In our test case NIVAs researchers suggested a range of alternative water levels to be evaluated.

(2) **Experts identify impacts and their indicators.** Identification of river ecological and user interests for which impacts are to be evaluated, and identification of the impact indicators.

(3) **Experts identify critical river reaches and time periods.** In order to identify critical time periods, optimal water level curves for each ecological value and user interest chosen must be identified.

(4) **Experts construct pressure-impact curves.** 'Pressure- impact' curves for each of the ecosystem functions and services of interest summarize expert knowledge of the dose-response linkages. It is a special interpretation of so-called value functions used in multiple criteria analysis. A preliminary step to facilitate expert judgment is conversion of stream-flow to water level and water surface extent curves.

(5) **Consultant inputs alternative-impacts matrix for critical reaches and periods.** This step involvesthe analyst 'translating' the expert 'hand-drawn' pressure impact curves to a format that can be used in the MCA software tool.
(6) **Consultant generates multi-criteria ranking using software.** The analyst uses the DEFINITE software tool to generate a multi-criteria optimal water level curve for each 'critical period' and river reach. The baseline assessment is based on equal weighting of all ecosystem functions and services.
(7) **Stakeholders compare importance of impacts.** Stakeholders' evaluation can weigh the relative importance of impacts differently, if desired. Alternative preferences for different ecosystem services may lead to the identification of different optimal water level curves and environmental flow regimes.
(8) **Hydropower producers calculate production level at different flow alternatives.**
(9) **Stakeholders/policy-makers weigh cost of foregone hydropower against optimal multiple use environmental flow alternatives.** Optimal multi-criteria water level curves above do not considered costs to hydropower until the final step of the analysis; PIMCEFA has been adapted to consider hydropower's national strategic importance and policy-makers preference to evaluate hydropower opportunity costs explicitly against all other criteria.

## 4.4 THE APPLICATION OF PIMCEFA TO ØYEREN

PIMCEFA was tested in the Øyeren River delta in the Glomma and Laagen River basin in 2007–2008 at the Øyeren and Høyegga sites, involving researchers and well as local experts from research and local water user interests. The evaluation was part of the EC project STRIVER (www.striver.no), which included testing other IWRM methodologies in other focus areas throughout the Glomma and Laagen River basin (see Chapter 2, Nesheim *et al.* of this volume).

### 4.4.1 Stakeholders identify range of alternatives

The first step in any decision analysis is to define the alternatives to be evaluated. Berge *et al.* (2002) conducted an evaluation of different water level regulation alternatives for the Øyeren delta through a series of environmental impact studies between 1994 and 2000. Four different water level regulation scenarios were identified by a steering committee composed of authorities, hydropower and researchers. The water level curves of the four alternative regulation regimes are shown in Figure 4.2:

*Alternative 1:* Before Bingsfoss Hydropower Plant ("FørBingsfoss"). Water level regulation practiced before 1934 when Bingfoss HP was built, and during the period 1978–1995.
*Alternative 2:* After Bingsfoss Hydropower Plant ('EtterBingsfoss'). Water level regulation practiced during the period 1945–1978
*Alternative 3:* Trial water level regime ('prøvereglement'). Test water regime from 1995–2000.
*Alternative 4:* Bird protection alternative ('Fuglealternativ'). Water level regime identified as most preferable for migratory birds based on observations of 'favorable' years for bird populations.

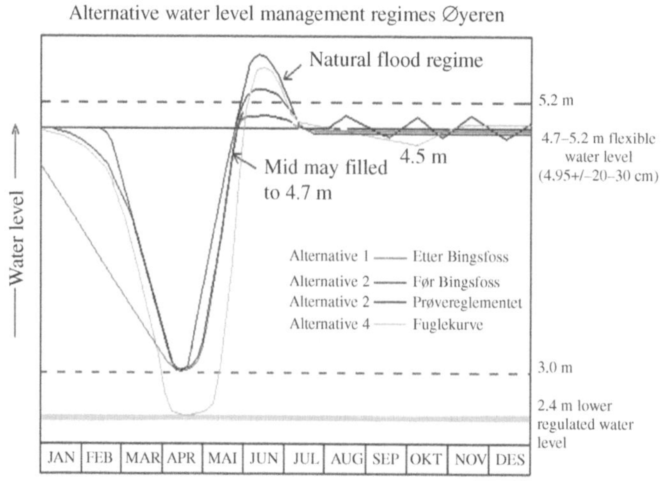


**Figure 4.2** Identification of alternative water level regimes. Source: based on Berge *et al.* (2002).

As we will see, the PIMCEFA approach can be used to identify an optimal water level for a given critical period and river stretch, by looking at a wider set of alternative water levels. In this chapter we will evaluate which of the alternatives best matches the results of PIMCEFA. In PIMCEFA the choice of alternatives need not be restricted to any specific stream flow alternatives. The analysis discussed below is therefore not restricted to the four alternatives evaluated by Berge *et al.* (2002). However, as a minimum the range of stream flow /water levels considered relevant by decision-makers must be determined. In our study this was the same as considered by Berge *et al.* (2002): 2.4 meters – 6.4 meters relative to a reference station.

## 4.4.2 Experts identify impacts and their indicators

The ecological and user values that were considered in the Øyeren test case were:

Ecological values:

- Fish
  - Spawning pike
- Birds
  - Ruff
  - Teal
- Macrophytes

Use values:

- Boating
  - 2.5–5 meters
  - 5–6.4 meters
- Farming
- Hydropower production

This list was based to a large extent on the availability of experts at short notice on the most relevant ecological and use values. The list is therefore not as exhaustive as it would be in a full evaluation of river regulation. Figure 4.3 shows a screenshot of the effects hierarchy in DEFINITE. Analyses were carried out with and without hydropower considered alongside other ecological and use values. In the account provided below our discussion focuses on how optimal water levels for all other uses impact on hydropower in a separate analysis outside the DEFINITE software. There are several reasons for these last two steps:

- hydropower regulation of the catchment is the upstream driver regulating water level;
- hydropower is a national strategic interest – authorities prefer to see an explicit evaluation of trade-offs between optimal water levels for other uses and resulting opportunity costs to hydropower (allowing for comparisons of costs of environmental flow across catchments); and
- optimal water levels initially defined independently from hydropower provide a stand-alone definition of ‘environmental flow’, before hydropower compensation considerations are considered.

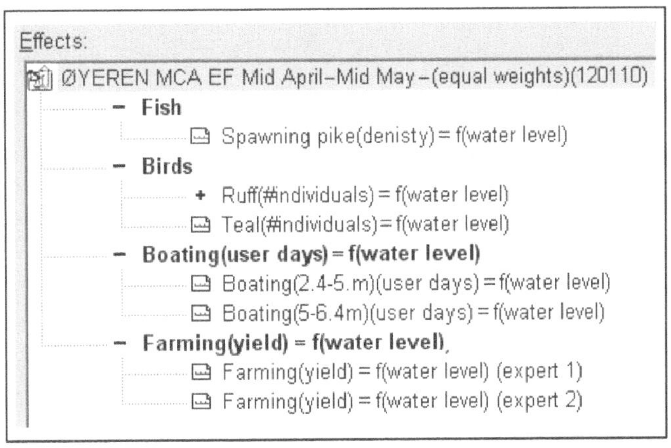


**Figure 4.3** Hierarchy of impact indicators identified by experts. Source: DEFINITE software screenshot.

### 4.4.3 Experts identify critical river reaches and time periods

In the case of Øyeren this step could be skipped with the expert group because Berge *et al.* (2002) had already evaluated optimal water level curves for different interests (e.g., protection of migratory birds). Optimal water level curves (Figure 4.2) establish the maximum or minimum water levels that define the best possible situation across seasons for the ecological/use component in question. This helps to determine the impact range that is to be characterized in the 'pressure impact' curves step. In Øyeren experts decided to distinguish impacts according to three different critical periods:

- April–May
- May–July
- August–September.

In this chapter we present results for the period April–May. A further assumption was that impacts of water levels in any particular year would be considered for up to a period of five years after that. This was meant to assure that cumulative inter-annual impacts of water levels in any particular year would be considered by experts. In practice it was observed that most of the impact would occur in the same year as the water levels being considered.

Impact indicators may be specific to the 'critical periods' being considered. Figure 4.3 shows the indicators that were used for the period of mid-April to mid-May. Notice that e.g. macrophyte vegetation is not considered in this period as it is largely outside the vegetation growth season in this area. Figure 4.3 also shows some particularities of the PIMCEFA method using the MCA software DEFINITE.

It is possible to handle differences in expert opinion on a particular pressure impact curve by creating individual indicators for each expert opinion (e.g., expert 1 and expert 2). In principle, different expert opinions are given equal weight, but the MCA software also allows the weighting of individual expert opinion. Figure 4.3 also illustrates that separate indicators must be specified for the increasing and decreasing portions of a pressure-impact curve, because the software algorithm cannot handle higher order exponential curves that are not monotonically increasing or decreasing. This is illustrated for 'boating' which has an optimal water level at five meters.

Figure 4.4 shows the basic structure of the impact matrix *before* pressure-impact curves have been evaluated by the software. The right column illustrates the hierarchy of impact criteria shown in Figure 4.3. C/B shows whether the impact curve is increasing or decreasing as a function of increasing water level. Minimum and maximum range of the pressure impact curves are indicated in the next columns.

| | | C/B | Standardization method | Minimum Range | Maximum Range | 2.0 | 2.4(LRV) | 2.5 | 3.0 | 3.2 | 3.5 | 4.0 | 4.5 | 4.6 | 4.8 | 5.0(HRV) | 5.2 | 5.5 | 6.4 |
|---|---|---|---|---|---|---|---|---|---|---|---|---|---|---|---|---|---|---|---|
| – | Mid April - Mid May | | | | | | | | | | | | | | | | | | |
| – | Fish | | | | | | | | | | | | | | | | | | |
| | Spawning pike(denisty)=f(water level) | + | free form | 4.50 | 5.00 | 2.00 | 2.40 | 2.50 | 3.00 | 3.20 | 3.50 | 4.00 | 4.50 | 4.60 | 4.80 | 5.00 | 5.20 | 5.50 | 6.40 |
| – | Birds | | | | | | | | | | | | | | | | | | |
| – | Ruff(#individuals)=f(water level) | | | | | | | | | | | | | | | | | | |
| | Ruff(#individuals)=f(water level)(expert 1) | ● | free form | 2.40 | 6.00 | 2.00 | 2.40 | 2.50 | 3.00 | 3.20 | 3.50 | 4.00 | 4.50 | 4.60 | 4.80 | 5.00 | 5.20 | 5.50 | 6.40 |
| | Ruff(#individuals)=f(water level)(expert 2) | ● | free form | 2.40 | 5.00 | 2.00 | 2.40 | 2.50 | 3.00 | 3.20 | 3.50 | 4.00 | 4.50 | 4.60 | 4.80 | 5.00 | 5.20 | 5.50 | 6.40 |
| | Teal(#individuals)=f(water level) | ● | free form | 2.40 | 6.40 | 2.00 | 2.40 | 2.50 | 3.00 | 3.20 | 3.50 | 4.00 | 4.50 | 4.60 | 4.80 | 5.00 | 5.20 | 5.50 | 6.40 |
| – | Use values | | | | | | | | | | | | | | | | | | |
| – | Boating(user days)=f(water level) | | | | | | | | | | | | | | | | | | |
| | Boating(2.4-5.m)(user days)=f(water level) | + | free form | 2.40 | 5.00 | 2.00 | 2.40 | 2.50 | 3.00 | 3.20 | 3.50 | 4.00 | 4.50 | 4.60 | 4.80 | 5.00 | 5.20 | 5.50 | 6.40 |
| | Boating(5-6.4m)(user days)=f(water level) | ● | free form | 5.00 | 6.40 | 2.00 | 2.40 | 2.50 | 3.00 | 3.20 | 3.50 | 4.00 | 4.50 | 4.60 | 4.80 | 5.00 | 5.20 | 5.50 | 6.40 |
| – | Farming(yield)(=f(water level) | | | | | | | | | | | | | | | | | | |
| | Farming(yield)(=f(water level) (expert 1) | ● | free form | 2.40 | 5.60 | 2.00 | 2.40 | 2.50 | 3.00 | 3.20 | 3.50 | 4.00 | 4.50 | 4.60 | 4.80 | 5.00 | 5.20 | 5.50 | 6.40 |
| | Farming(yield)(=f(water level) (expert 2) | ● | free form | 4.80 | 7.00 | 2.00 | 2.40 | 2.50 | 3.00 | 3.20 | 3.50 | 4.00 | 4.50 | 4.60 | 4.80 | 5.00 | 5.20 | 5.50 | 6.40 |
| | Hydropower(GWh)=f(water level) | ● | free form | 3.20 | 4.80 | 2.00 | 2.40 | 2.50 | 3.00 | 3.20 | 3.50 | 4.00 | 4.50 | 4.60 | 4.80 | 5.00 | 5.20 | 5.50 | 6.40 |

**Figure 4.4** Impact matrix for critical period of mid April–Mid May (DEFINITE software screenshot). Note: Impacts are simply the different water levels of each alternative. Water levels are converted to standardized impacts using 'pressure-impact' curves.

In MCA alternatives (columns) are commonly represented as policy/project alternatives. Alternatives are subject to an impact assessment (IA) 'outside' the MCA. The results of the IA for each criterion would be entered into the table below. In PIMCEFA, stream flow/water levels are proxy variables for the different alternative management regimes. The principle issues in an IA can then be documented using pressure-impact curves to replace 'marginal value functions'. The numbers in the impact matrix are now simply the different water levels that each impact is to be assessed at (using the pressure impact curves). The number of alternative water levels to be considered can be determined by stakeholders if the alternatives are to be limited to only those that are of policy-relevance. If the analysis is 'prospective' experts can propose a range of water levels they wish to rank. Each water level considered in Figure 4.4 simply reflects a collection of all the water levels that experts expressed an interest in when considering their individual pressure impact curves.

### 4.4.4 Experts construct pressure-impact curves

'Pressure impact curves' – a term we have adopted in the context of environmental flow – are similar, but not identical, to marginal value functions in the multiple criteria analysis literature (Beinat, 1997; Munda, 2008). Figure 4.5 shows the simplest of pressure impact curves, with water level indicated on the horizontal axis and impact indicated on the vertical axis. No impact relative to the optimal situation is given a score of 1 or 100%. Full loss of the ecological or use value is given a score of 0%. Note that pressure impact curves need not be specified across the full range of water levels considered in the analysis – pressure impact curves capture the water level 'niche' at which the ecological component/use value is to be found.

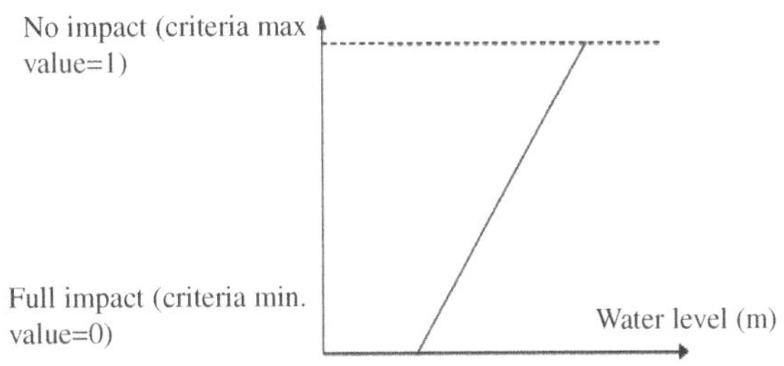


**Figure 4.5** The most basic form of a pressure impact curve (linear increasing impact with reduction in water level).

In PIMCEFA experts are asked to give 'impact' on the vertical axis a specific biophysical interpretation as the range between total loss and optimal conditions of the ecological or use value in question. In MCA literature the marginal value functions interprets the vertical axis as a dimensionless 'value' (Beinat, 1997) or 'attractiveness' (Munda, 2008). We argue that a 'pressure-impact' interpretation conforms more to a dose-response relationship between two observable phenomena, making it possible to validate expert judgment with models or empirical data. It avoids the confounding of dose-response with 'attractiveness' and 'value' judgments. All value judgments of (relative) importance are handled in the later step of weighting. This also provides a clearer distinction between the role of the technical expert and decision-maker who must weigh interests against each other.

Figure 4.6 shows a hand-drawn example of a pressure impact curve produced during the expert panel meeting. In the illustration for the bird species Ruff (Norwegian 'Brushane'), the judgments of various experts have been compiled in the same drawing after comparing curves that were drawn separately. They show some discrepancy between experts.

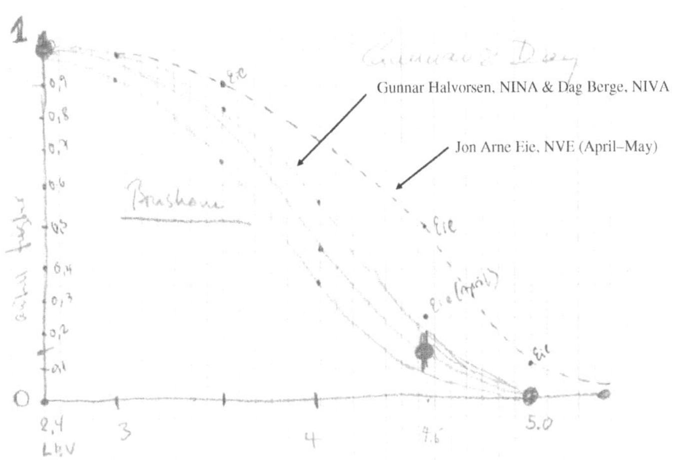


**Figure 4.6** Expertjudgment is recorded as individual 'pressure-impact' curves then compared (three experts on the Øyeren wetland bird species 'Ruff').

### 4.4.5 Consultant inputs alternative-impacts matrix for critical reaches and periods

The analyst next converts hand drawn pressure-impact curves to numerical scores for each criteria. In DEFINITE input of marginal value functions is called 'standardization of effects' (Jansen *et al.*, 2003). Figure 4.7 illustrates how the hand-drawn pressure impact curves are input into the DEFINITE MCA software. The curves convert water levels to impacts for each criterion on a scale from 0 to 1. The standardized impact scores of different water levels are shown along the bottom row of the screenshot. Standardized impact scores of different water levels are shown along the bottom row of the screenshot in Figure 4.7. These impact scores become part of the 'impact matrix' illustrated in Figure 4.4. A number of **functional forms** can be selected in DEFINITE (e.g. linear, convex, concave, S-shaped, free-form). In our opinion this is one of the features that determined our choice of DEFINITE as software for PIMCEFA. Other MCA software (e.g., Expert Choice) only allows linear convex and concave curves, making it impossible to use expert free-form hand-drawn pressure-impact curves. One limitation of standardization procedure in DEFINITE is that pressure-impact curves must be monotonically increasing or decreasing, making it impossible to use functions with local peaks (maxima) or troughs (minima). This can be overcome by splitting the criteria into two (or more) standardization functions representing increasing and decreasing sections of the pressure-impact curve. This is illustrated in Figure 4.9 (next page), where riparian 'macro-vegetation' are eliminated below and above water levels of 2 meters and 5.5 meters respectively, with a peak at a water level of 4 meters representing optimal growth conditions (May–July).

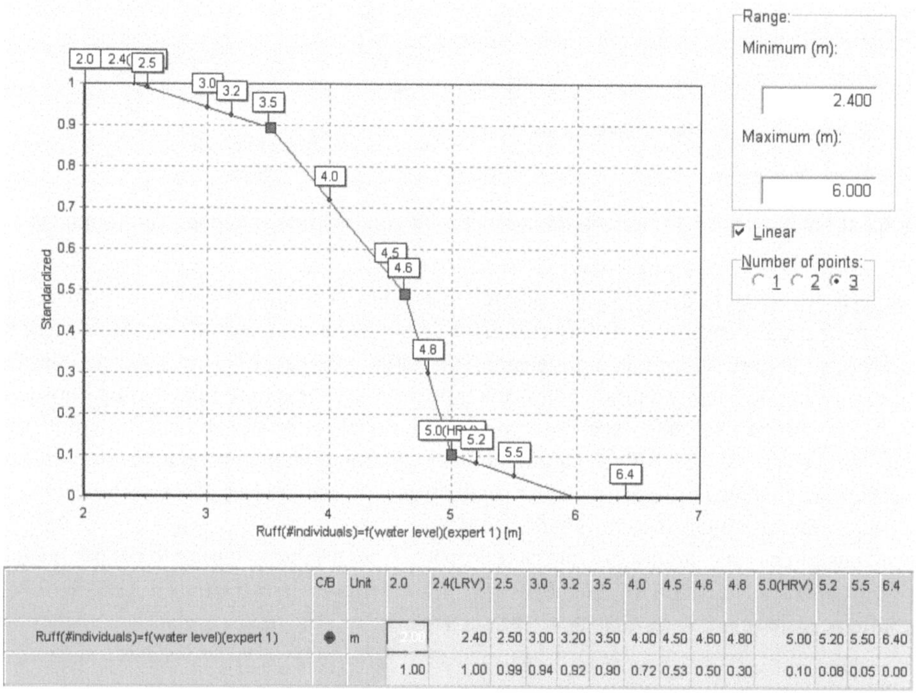


**Figure 4.7** A single "pressure-impact curve" for Ruff bird species as input to the multiple criteria analysis tool.

### 4.4.6 Dealing with uncertain expert judgment

The discrepancy "between expert opinions, shown as alternative pressure-impact curves in Figure 4.6 above, illustrates one type of scientific or technical uncertainty regarding dose-response. Similarly, a single expert might not feel comfortable with being asked to specify a single curve, but may accept specifying an uncertainty region, enveloped by a 'maximum' and 'minimum' likely impact curve. In the case of disagreement between two experts, we handled this by inputting the 'lowest' and 'highest' impact curves to the DEFINITE software. The opinion of each expert can then be weighted by the decision-maker in a separate exercise (in the analysis below experts were by default given equal weight). Similar curves were drawn for the farming impact indicator where there was discrepancy in expert judgment.

With experts willing to give more information regarding the nature of technical/scientific uncertainty other modeling techniques in the MFA are conceivable. When experts base their judgment on statistics or model simulations median, 10% and 90% confidence curves can be specified (Figure 4.8). Different curves could then be weighted according to the confidence levels or marginal probabilities of each curve. However, care must be taken not to overburden the intuitive simplicity of 'pressure-impact' curves with statistical interpretations. In addition, probability based weighting requires further thinking as confidence levels express probabilities of exceeding test values (or curves), rather than the marginal probability of the test value occurring. Marginal probabilities of two test values on the other hand do not sum to 100%. If dose-response relationships can be modeled expert opinion will probably be based on the mean or median of model predictions. DEFINITE software also contains the ENVALUE procedure (Beinat, 1997) which offers an iterative approach to specifying marginal value functions simultaneously with determination of relative weights of criteria. ENVALUES includes three different elicitation steps in DEFINITE (Janssen *et al.*, 2003):

(1) Specification of 'value regions' (see lower two panels of Figure 4.9);
(2) Ranking of effect criteria; and
(3) Indirect assessment of alternative water levels in terms of pairs of criteria.

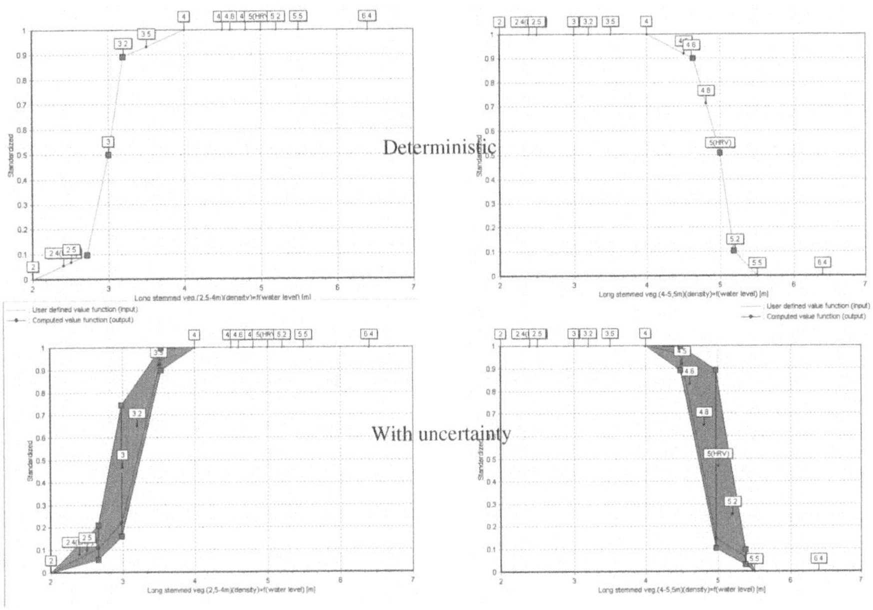


**Figure 4.8** Individual expert uncertainty about pressure-impact can also be captured as 'confidence intervals', as opposed to a single deterministic curve ('pressure-impact'curves for long stemmed wetland macrovegetation).

These three steps allow joint specification of marginal value functions and relative weights of criteria, and can be repeated to refine judgments of both. While the ENVALUES procedure has the advantage of helping experts and decision-makers describe their uncertainty, its drawback for PIMCEFA is that it requires the same persons to simultaneously play the roles of technical experts and decision-makers; mixing the tasks of specifying pressure-impact/value functions and specifying weights, respectively. We argue that a particular strength of PIMCEFA is a clearer separation of these roles.

## 4.4.7 Consultant generates multi-criteria ranking using software

In this step the analyst uses the DEFINITE software tool to generate a multi-criteria optimal water level curve for each 'critical period' and river reach. The baseline assessment is based on equal weighting of all ecosystem functions and services. Figure 4.9 compares the alternative water management regimes identified in Berge *et al.* (2002) with the multi-criteria optimal water levels determined using PIMCEFA for two critical periods. Aggregate weighted scores of alternative water levels for April–May and July–September are shown in the lower two panels. The water level management regimes evaluated by Berge *et al.* (2002) are shown in the upper panel. Due to the discretisation of water level intervals used in the analysis, PIMCEFA identifies an optimal water level interval (lower panels). Policy-relevant interpretation of these results can be made for even this simple example. PIMCEFA highlights a potential high conflict of interest between groups of ecological and user interests in the April-May critical period, represented by two separate local optimal water levels. In July-September there is a single optimal water level (peak), suggesting that conflict of interest is low. Regarding uncertainty in expert judgment of 'pressure-impact' curves, this is reflected in a number of aggregate scores for alternative water levels being close to the score of the optimal water levels (peak values in the lower panel diagrams). The optimal water levels determined independently using PIMCEFA coincide with most of the scenarios identified in Berge *et al.* (2002), providing some independent validation of the method.

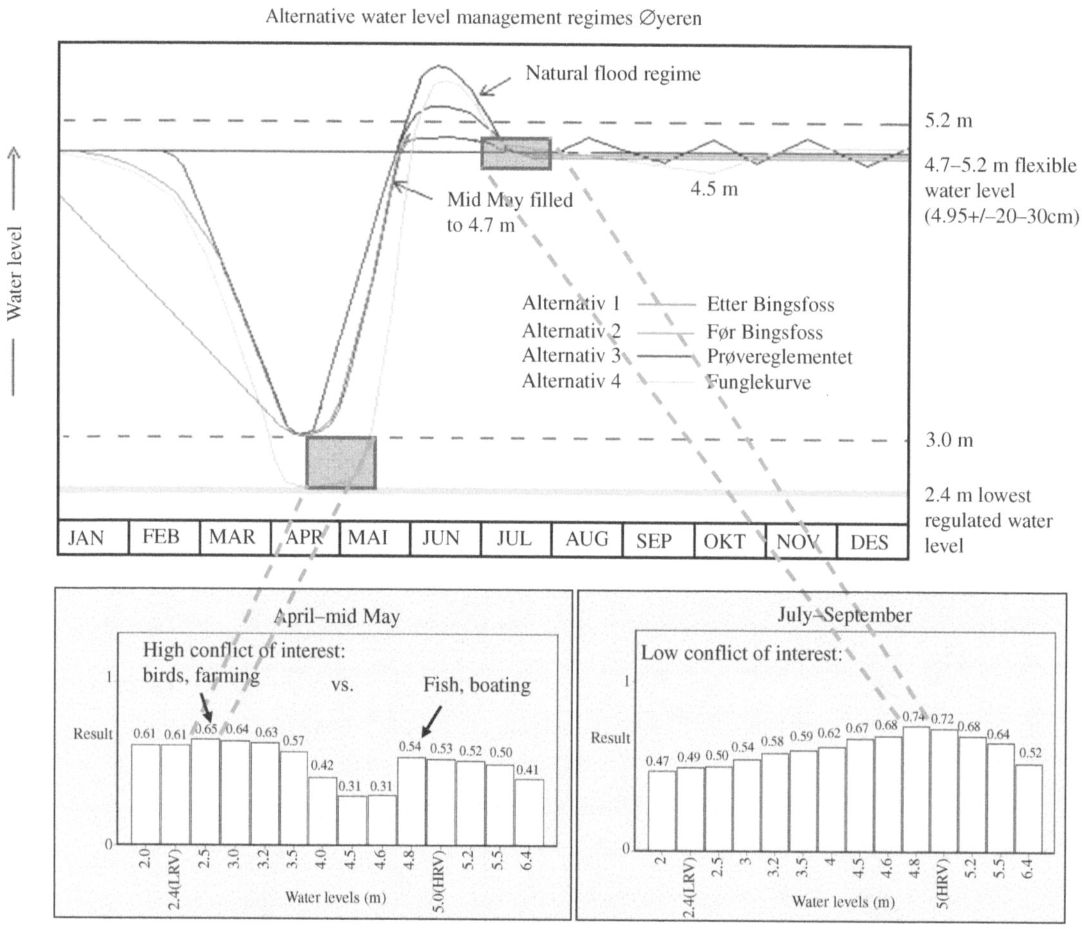


**Figure 4.9** Comparison of Berge *et al.* (2002) alternative regimes with PIMCEFA multi-criteria optimal water levels. Aggregate weighted scores of alternative water levels for April-May and July-September as determined in DEFINITE (lower panels). The optimal water levels determined independently using PIMCEFA coincide with most of the scenarios identified in Berge *et al.* (2002). In the analysis results illustrated above hydropower was included in the analysis.

## 4.4.8 Stakeholders compare importance of impacts

The last step of the MCA analysis involves evaluation with stakeholders. The simplest form of evaluation would involve stakeholders validating the conclusions of the analysis based on equal weights given to different ecosystem services. If stakeholders agree with the pressure-impact curves and with the resource conflict illustrated by the optimal water level scores generated in PIMCEFA, the next step would be to separate the stakeholders into 'interest groupings'. Each interest grouping would conduct a separate weighting session. The MCA software would then be used to input weights to different interests. Several outcomes of a weighting exercise with stakeholders are possible:

- the weighting of stakeholders strengthens the divergence in optimal water levels for (April–May) identified by experts using equal weighting of pressure-impact curves;
- the weighting leads to a convergence of optimal water levels – this would show that the resource conflict is not as great a 'political' problem as that suggested by the impact analysis conducted by the experts; and
- the weighting does not modify the conclusions of the analysis. This might be the case where pressure-impact curves had been specified for a number of very limited 'niche' ranges of water levels. In such cases, weighting of interests will have little impact on the outcome, showing that there is little room for trade-offs between interests.

## 4.4.9 Hydropower producers calculate production level at different flow alternatives

Figure 4.10 shows an estimation of hydropower production (GWh) for the period April-May at different water levels evaluated in the MCA software. Hydropower losses at lower water levels are due to reductions in production regulation flexibility when water levels are 'fixed' at levels specified by other interests. In the April-May example, the option to regulate water levels is foregone once a year, while water levels being fixed in the autumn may lead to several foregone opportunities to regulate stream flow into and out of the Øyeren reservoir. In the case of Øyeren the hydropower producer GLB has included foregone power due to this reduced flexibility, as well as reduced effect at the Solbergfoss and Bingsfoss hydropower stations due to lower water levels. The difference between a regime that favors birds and farming versus one that favors boating and fish is equivalent to incremental hydropower production of about 30 GWh or approx. NOK 7.5 million (0.25 NOK/kWh).

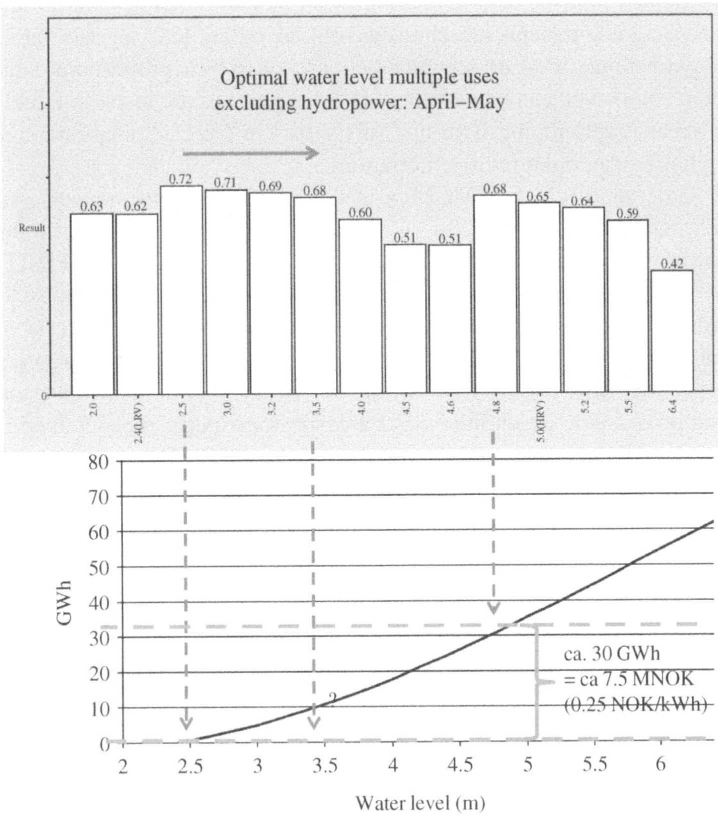


**Figure 4.10** Estimation of hydropower production at different water levels and comparison with optimal water levels for other users. Optimal water level scores shown in the upper panel were evaluated in DEFINITE not including hydropower (all other interests).

### 4.4.10 Stakeholders/policy-makers weigh cost of foregone hydropower against optimal multiple use environmental flow alternatives

The comparison of the multiple criteria water levels and hydropower production makes it easier to identify 'win–win' situations where impacts on other water users can be mitigated, or compensated. Only certain water level changes may impact on a specific water user or ecological value. These impacts may be mitigated using technical measures, or if not there is a potential for using part of income from additional hydropower production for financial compensation. It may not be possible to identify such 'win–win' or compensation opportunities in many cases that will reduce conflicts of interest. However, the use of 'pressure-impact' curves together with multiple criteria analysis early on in an Impact Assessment will help identify where conflicts of interest are due to uncertainty about impacts, and where they are due to real differences in preferences between stakeholders.

## 4.5 OTHER MULTI CRITERIA APPROACHES TO ESTABLISHING ENVIRONMENTAL FLOW

This section highlights some of the principal differences between the PIMCEFA approach to establishing environmental flows and other multiple criteria analysis approaches which we are aware of within this field. The review and comparisons are not exhaustive, but illustrative of the kinds of decision situation in which PIMCEFA could play a role.

The DRIFT approach by Brown *et al.* (2003, 2006) and – Downstream Response to Imposed Flow Transformation; and Reichert *et al.* (2007) use of causal probabilistic networks are looked at here. Environmental flow has traditions in hydraulic, hydrological and habitat modeling (Berge *et al.*, 2010). DRIFT is a holistic ecosystem modeling method addressing five routine ecosystem components – fluvial geomorphology, water quality, plants, aquatic invertebrates and fish. TheDRIFT multi-criteria analysis tool is one module within the holistic DRIFT approach developed by King *et al.* (2003). The DRIFT MCA computes 'river integrity scores' for different stream flow levels using a multi-attribute value function of the ecosystem components, (Brown *et al.*, 2003). DRIFT characterizes ecological impacts using ten flow classes that together describe dry/wet season; intra-annual flooding severity and return period of floods. For each flow class the ecosystem components are characterized according to a severity rating with six levels ranging from 'no change' to 'local extinction' (loss) or 'abundance increase to pest proportions >500%'(gain). The relative importance of each ecosystem component can be weighted. Database entries are made in Excel. The river integrity score is computed using integer linear programming with the Solver tool in Excel. Social and economic consequences are addressed in a subsequent phase outside the multi-criteria tool.

DRIFTs MCA uses the same approach as PIMCEFA in describing 'pressure-impact' curves where impact has a verifiable biophysical interpretation (converted to a score), rather than a dimensionless rating. DRIFT uses predetermined flow and severity levels, whereas the 'standardization curves' in DEFINITE allow any number of levels to be specified. The interpretation of 'severity' in DRIFT is fixed, whereas it must be flexible/generic in the PIMCEFA approach in order to accommodate impacts on socio-economic user interests, in addition to ecological components. The greatest difference is that DRIFTs MCA is a dedicated tool for assessing ecological impacts in rivers, whereas DEFINITE is a generic MCA tool. One of the advantages of a dedicated tool such as DRIFT is a standardization of evaluation procedures, which may make it easier for experts to conduct consistent evaluations across different river types. On the other hand, a generic MCA tool such as DEFININTE can more easily be adapted to different types of water resources such as rivers, deltas, lakes. For example, in the multiple-use context of South Africa, DRIFT evaluates social and economic impacts of water users outside the MCA model in 'scenario creation' and 'economics' modules. In the PIMCEFA example above, only consequences for hydropower were handled outside the analysis, due to the special nature of hydropower as the main driver of stream flow. To the extent that quantification of the economic damage to different users is to be made explicit in the analysis, and differential weighting of social-economic interests are not needed, it can be convenient to analyze socio-economic impacts of flow scenarios outside MCA such as in DRIFT. Finally, a slight technical advantage of DRIFT may be that it is run entirely within an Excel environment; DEFINITE is astand-alone software with the possibility to import Excel files.

Reichert *et al.* (2007) used probability networks to assess decisions regarding river rehabilitation in Switzerland. This is a hybrid approach, where probability networks integrate the results of simulation models, expert opinion and multiple criteria weighting of the importance of impacts. The authors designed a model to evaluate river rehabilitation options affecting river morphology and hydrology, impacting on a number of decision criteria: fish, shoreline communities, vegetation, benthic organisms, and socioeconomic consequences. The advantage of a causal probability network is that causal relationships between physical, ecological and socioeconomic components can be made explicit. In multi-criteria analysis using additive value functions each criteria – and it value functions – are assumed to be independent (while it is possible to include simple linear correlations between impacts in DEFINITE). Because causal networks are generic, ranking of alternatives can be based on multi-attribute value functions using relative weights, such as in MCA, or other approaches to describing preferences such as utility functions (Reichert *et al.*, 2007) – utility functions can capture attitudes to risk, but also require additional preference elicitation procedures to those used in traditional MCA. Furthermore, causal probability networks can be used to model dose-response relationships (Beinat, 1997) explicitly. In PIMCEFA, 'pressure-impact' curves summarize all dose-response relationships between stream flow and the impact of decision-making interest in a single curve. When experts view this as a 'leap of faith' and have specific models or data they would like to include explicitly in the decision-analysis, causal probability networks can be a meta-modeling tool to complement MCA (Barton *et al.*, 2008). Finally, as their name suggests, probability networks provide an explicit and consistent way of dealing with uncertainty in expert judgment and model predictions.

On a scale of specific to generic application, we would rank the approaches as (1) DRIFT (2) PIMCEFA (3) Causal probability networks. We also sense that the generic nature of multi-criteria analysis software, and – even more so – causal probabilistic networks, comes at the price of increasing information demands on experts. From our brief review above we think the approaches are not mutually exclusive; elements from each can be combined as the information available and decision-making situation dictates.

## 4.6 CONCLUSIONS – POSSIBLE IMPROVEMENTS IN THE METHODOLOGY

A number of improvements to the PIMCEFA method are summarized below based on the previous two sections.

Undocumented impacts need to be considered using a '**precautionary principle**'. Undocumented future impacts may be included as separate criteria with their own pressure-impact curves. They could be weighted in order to force the optimal water level solution up or down depending on the expected nature of the impacts that were not included. While subjective, the role of precaution and risk attitudes of stakeholders would therefore be explicitly documented in the analysis, and open for revisions as documentation became available. Alternative approaches to modeling risk attitudes include the use of utility functions rather than value functions (Reichert *et al.*, 2007).The Øyeren example considers the impact of different water levels in a given period on different ecological and use values. The impact of seasonal and inter-annual variation flows is not captured by the methodology. For seasonal variation pressure-impact curves could conceivably be constructed by experts where the indicator of pressure is different in maximum and minimum flows between seasons (i.e., pressure is now measured as change in water level). DRIFT has an explicit characterization of the severity of inter-annual flooding and return periods for each ecological component (Brown *et al.*, 2003) – this approach could be adopted in PIMCEFA.

Given expert specialization there is a tendency to focus on indicators of species or specific user interests. Part of the expert evaluation should also be dedicated to finding system indicatorswhich may also be good 'proxies' for (or may replace) species specific indicators. Some examples from Øyeren wetlands were suggested by experts involved in the evaluation, for instance:

- helophyte zone as primary production supplier
- turbidity
- keystone species.

Red list species may be over-represented in the analysis simply because of greater access to documentation. By using a criteria hierarchy (e.g. birds = >red list species) all information can be used without the number of species itself determining the weight of birds in the multiple criteria analysis (as compared to other ecological elements or user interests).

Various approaches to dealing with uncertainty in pressure impact curves are available (confidence intervals, minimum and maximum curves, weighted average of different expert curves). The use of confidence intervals was explained to experts, but very few opted to express uncertainty, choosing instead a single deterministic curve. Uncertainty in expert judgment was more easily captured when evaluating differences between expert opinion (max and min estimates). Confidence intervals may be more relevant to the individual expert if asked to specify a pressure-impact curve over a range of background conditions (e.g., low seasonal variation in water levels versus high variation). A causal probability networks would assist experts in integrating background conditions for 'pressure-impact' curves in their assessments (Reichert *et al.*, 2007).

Evaluate the correlation of pressure-impact curves before conducting the multi-criteria weighting. This may significantly simplify the analysis as highly correlated impacts may mean that multiple impacts can be represented with a single "proxy" pressure-impact curve. For example, ecological and use values relying on surface area at given water depths are highly correlated with a bathymetric curve. A bathymetric curve can be used to construct water and mud-flat surface area curves, as well as indices of area-depth diversity, and temperature-depth diversity which may be good proxies for environmental niches and biodiversity. If correlation of impacts is non-linear, causal probability networks can be used to specify the correlation structure.

Economic uses may depend on access points to the water body at given water depths. Maps should be provided as background documentation for experts at the start of the meeting for relevant 'critical' river sections.

Given the limitation on fixing the pressure-impact curve to three points, a good rule of thumb may be to ask experts to provide the 10%, 50% and 90% impact water levels, in addition to the end points. Our experience from the expert panels used to elaborate pressure-impact curves indicates that experts do not have the data or models to evaluate impacts with any more precision than this (even in the well documented Øyeren case).

'Critical periods' can also be differentiated by flood years and normal years to capture the importance of flood related processes. Again, DRIFT methodology explicitly evaluates the impact on ecological components under flood conditions (Brown *et al.*, 2003).

Critical impact curves may be defined for both critical periods and water body sections in order to better capture temporal and spatial variation. In the present example only temporal variation was accounted for. To the extent possible, 'expert/scientific' and 'political/stakeholder' roles should be separated in the MCA process. Definition of pressure impact curves should be easier to replicate in a separate expert group if the mixing of roles is minimized. The task of fixing pressure-impact curves for user interests mixes the role of expert and stakeholder. Further care has to be taken in trying to separate these roles as far as possible in the expert evaluation process, despite the fact that people have multiple professional and private roles in everyday life. One possible improvement on our case study example would be to bring in non-local representatives of the user interest as expertise for the pressure-impact curve definition (with the disadvantage that they will know less about local conditions).

As suggested above several of the improvements needed for PIMCEFA can be 'borrowed' from modeling approaches in DRIFT or causal probability networks. Other problems are more fundamental due to lack of data. However, expert and stakeholder willingness to have their assessments and judgments made public and exposed to scrutiny may be the biggest limitations to the implementation of multiple criteria analysis for evaluating environmental flow in Norway. Regarding experts, PIMCEFA integrates and summarized a lot of their expert knowledge, and may be seen as a threat in that their expertise is 'captured' and may be used by others. A system of publication of expert opinions summarized in 'pressure-impact' curves may help to both recognize the source of expertise and expose it to public scrutiny. Regarding decision-makers, multi-criteria-analysis has a disruptive effect on the political economy of projects and policy. MCA documents where conflicts of interest may lie, both in relation to other stakeholders and in financial terms for specific stakeholders. Nevertheless, environmental impact assessment and public hearings are mandatory for large hydropower projects and major revisions of HP concessions. Viewed as an approach to jointly considering technical aspects of EIA with stakeholders' preferences as expressed in hearings – rather than as simply an MCA software application – PIMCEFA may offer some ideas for new approaches in Norwegian river basin management.

## 4.7 REFERENCES

Barton, D.N. and og Berge, D. (2008) Pressure-Impact Multi-Criteria Environmental Flow Analysis (PIMCEFA). STRIVER Technical Brief No.3.http://www.striver.no/diss_res/files/STRIVER_TB6.pdf

Barton, D.N., Saloranta, T., Moe, S.J., Eggestad, H.O. and Kuikka, S. (2008) Bayesian belief networks as a meta-modelling tool in integrated river basin management – Pros and cons in evaluating nutrient abatement decisions under uncertainty in a Norwegian river basin. *Ecological Economics* **66**, 91–104.

Beinat, E. (1997) *Value Functions for Environmental Management*. Kluwer, The Hague.

Berge, D., Bjørndalen, K., Brabrand, Å., Andersen, R., Dale, S., Bogen, J., Bønsnes, T.E., Martinsen, T., Elster, M., Rørslett, B., Halvorsen, G. and Sloreid, S.E. (2002) Environmental studies in Lake Øyeren 1994–2000. Main Report. (In Norwegian), AkershusFylkeskommune, Oslo.

Berge, D., Barton, D.N., Nhung, D.K. and Nesheim, I. (2010) Environmental Flow. In *Science, Policy and Stakeholders in Water Management – An Integrated Approach to River Basin Management*. G. Gooch and P. Stålnacke (eds) Earthscan, London.

Brown, C.A. and Joubert, A. (2003) Using multi-criteria analysis to develop environmental flow scenarios for rivers targeted for water resource management. *Water SA* **29**(4).

Brown, C.A, Pemberton, C., Berkhead, A., Bok, A., Boucher, C., Dollar, E., Harding, W., Kamish, W., King, J., Paxton, B. and S. Ractliffe (2006) In support of water-resource planning – highlighting key management issues using DRIFT: a case study. *Water SA* **32**(2).

Dunbar, M.J., Gustard, A., Acreman, M.C. and Elliott, C.R.N. 1998: Overseas approaches to the setting River Flow Objectives. Env. Agency R&D Technical report W6-161. Wallingford, UK. Institute of Hydrology: 83 pp.

Halleraker, J.H. and Harby, A. (2006) International methods for deciding environmental flow – which of these are aplicable in Norway? NVE Miljøbasertvannføring, Report 9 2006, 69 pp. ISBN 82-410-0584-9.

IUCN (2003) *Flow – The Essentials of Environmental Flows*. IUCN, Gland.

Janssen, R. and Herwijnen, M.V. 2007, *DEFINITE 3.1.A system to support decisions on a finite set of alternatives (Software package and user manual)* Institute for Environmental Studies (IVM), VrijeUniversiteit, Amsterdam.

Janssen, R., van Herwijnen, M. and Beinat, E. (2003) DEFINITE 3.0 case studies and user manual. Report number R-03(03 VrijeUniversiteit Amsterdam (IVM).

Jowett, I.G. (1997) Instream flow methods: a comparison of approaches. *Regulated Rivers: Research and Management* **13**, 115–127.

King, J.M., Brown, C.A. and Sabeth, H. (2003) A scenario-based holistic approach to environmental flow assessments for regulated rivers. *River Resources Applications* **19**(5–6), 619–640.

MEA (2005) *Millenium Ecosystem Assessment. Ecosystems and Human Well-Being: Biodiversity Synthesis*. World Resources Institute, Washington, DC.

Reichert P., Borsuk, M., Hostmann, M., Schweizer, S., Spörri, C., Tockner, K. and Truffer, B. (2007) Conpcets of decision support for river rehabilitation. *Environmental Modelling & Software* **22**, 188–201.

Tharme, R.E. (2003) A global perspective on environmental flow assessment: emerging trends in the development and application of environmental flow methodologies for rivers. *River Research and Applications* **19**, 397–441.

# Chapter 5

# IWRM status in the Tagus River basin

*A. Rieu-Clarke, D. Campbell, M.M. Portela, A. Betâmio de Almeida, S. Beguería, S.M. Vicente-Serrano, I. López-Moreno, N. Lana-Renault and J.M. García-Ruiz*

## 5.1 INTRODUCTION

The purpose of this chapter is to examine the status of integrated water resources management (IWRM) within the Tagus River basin. In so doing the chapter will provide a short description of the basin, prior to identifying the main challenges in managing the river. The chapter will then go on to examine the institutional structures that are in place to manage the waters of the Tagus. The conclusion will seek to ascertain some of the main challenges in implementing IWRM within the Tagus River basin.

## 5.2 THE TAGUS RIVER BASIN

The Tagus River (Figure 5.1), located in the central region of the Iberian Peninsula, rises in east-central Spain, in Sierra de Albarracin at an altitude of 1,590 m and flows from East to West more than 1009 km to Portugal, where it empties in the Atlantic Ocean near Lisbon (UNEP, 2004). It is the longest river on the Iberian Peninsula. The catchment area of the Tagus is 80,100 km$^2$, 69% of which is located in Spain and the remaining 31% is in Portugal. In terms of major land use types, the area under shrubs and *maquis* accounts for approximately 25%, forest around 16% and non irrigated arable lands around 26%. Irrigated land is only about 3.5%. Though the precipitation in the basin ranges from around 500 mm to about 1200 mm, the majority of the basin falls within the range of about 500 to 700 mm per year. It is navigable for about 160 km from its mouth, while dams harness its waters for irrigation and hydroelectric power, creating large artificial lakes. The lower and upper courses pass through deep gorges and are broken by waterfalls. There have been efforts to reforest the land surrounding the river. The river's main tributaries are the: Jarama, Alberche, Tietar, Alagon, Guadelia, Almonte and Salor Rivers in Spain and the Erges, Ponsul, Zezere, and Sorraia Rivers in Portugal.

## 5.3 WATER USES AND PRESSURES IN THE BASIN

Around 80% of water demand within the Tagus River basin is for agricultural purposes, with around 20% for drinking-water supply. Further uses include hydropower, wastewater disposal, recreation and basic ecological functions. In addition, part of the flow of the Tagus in Spain is diverted to the Segura basin, thus supplying 1.5 million people in Southern Spain with drinking water, providing irrigation and supporting the ecosystem in the La Mancha Nature Reserve (Beguería, 2008).

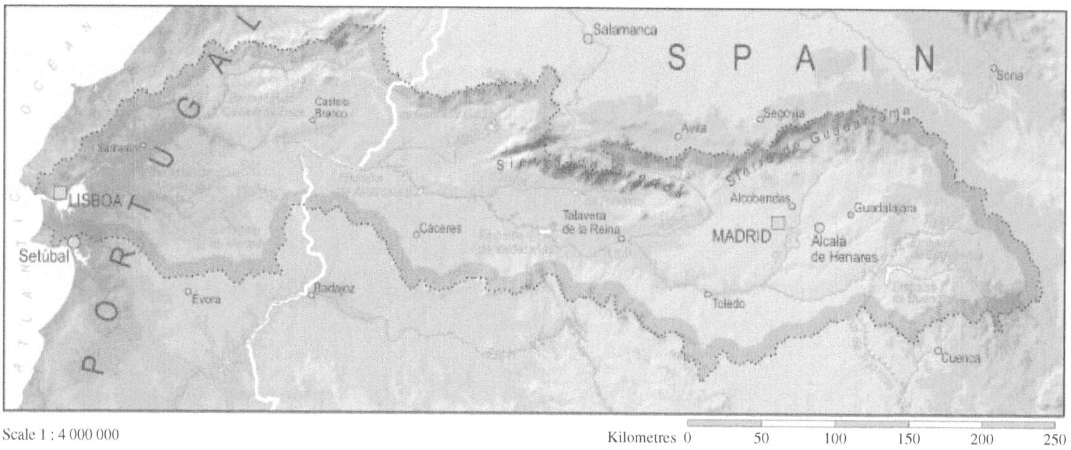


**Figure 5.1** The Tagus River basin (UNEP, 2004).

Within the Spanish part of the Tagus River basin, domestic water consumption, currently accounts for 1000 $hm^3$ per year; whereas in Portugal, domestic water consumption related to the Tagus accounts for 343 $hm^3$ per year. Two European capitals, Madrid and Lisbon, with a total of 11 million people depend on the river as their main source of urban water supply, as well as for agricultural purposes. The location of Madrid, Lisbon and other cities in the Tagus basin stresses the importance of urban uses of the water. In the case of Madrid there is a complex network of reservoirs and canals that ensure the water supply to more than 5.3 million inhabitants (Nesheim *et al.,* 2008).

The total storage capacity of the numerous reservoirs in the Tagus basin corresponds to 74% of the average annual runoff. In most of the basin natural flow conditions have now been replaced by regulated flow. The Tagus flows mostly through semi-arid lands, and government efforts have been dedicated to increasing land irrigation and creating hydroelectric power in its basin.

**Figure 5.2** The Goats Castle Dam, Tagus River, Portugal (Photograph: G.D. Gooch).

Major efforts to harness the Tagus and its tributaries for these purposes were undertaken from the 1960s, and by 1980 more than 60 dams had been built with a total installed power capacity of more than 1,200,000 kilowatts (Nesheim *et al.*, 2008).

The comparison between the river regimes before (1949–1969) and after (1969–1986) the construction of the Alcántara reservoir close to the border between Spain and Portugal show the consequences of river regulation. There has been a clear decrease of discharges from November to April, and an increase from July to September. There is an increasing demand of water for irrigation and urban supply. A total of 230,720 ha are irrigated (7.1% of the irrigated areas in Spain), and most probably this figure will increase in the future due to the opening of new private wells. The possibilities of getting more groundwater are still far away from the limits, since only 6.9% of the natural recharge is taken every year. Some of the reservoirs were constructed to store water for irrigation as is the case of many located in tributaries. The volumes of water for industrial waste are very low compared with other basins in North Spain (Nesheim *et al.*, 2008).

In Spain, the irrigation system within the Tagus River basin account for 2000 $hm^3$ per year, and the water transfer between the Tagus and Segura River accounts for 500 $hm^3$ per year; and therefore constitutes a major pressure on the resources of the Tagus River basin. In Portugal, water requirements for agriculture account for 1930 $hm^3$ on average (Beguería, 2008).

The transfer of water from the Tagus to the Segura River started in 1978 as a means of supplying water to irrigated areas in the South-east of Spain. The original plans estimated a volume of water transfer of about 1100 $hm^3$ per year. However, in practice the volume has only exceeded 500 $hm^3$ per year in exceptional circumstances. The volume of water transfer must be approved year by year (and even month by month, or season by season during certain conditions) by the Spanish Government, according to the volume of water stored in the Entrepeñas and Buendía reservoirs (Ministry of Environment, 2010).

In Spain, the industrial sector constitutes a further pressure on the resource; with the cooling of nuclear plants accounting for 1400 $hm^3$ per year. In Portugal, annual water requirements for industry account for 127 $hm^3$.

The Tagus estuary is one of the three most important in Europe and is the most important wetland in Portugal (Fernandes *et al.*, 2005). A part of the area is a nature reserve under national law and is also designated as an EU Special Protection Area for wild birds, covering some 40,000 ha (EC, 1979). In addition, the wetland is recognised as a site of international important under the Ramsar Convention (1971).

## 5.4 WATER CONFLICTS IN THE TAGUS BASIN

Largely due to the legal and institutional framework that has been established for the management of the Tagus River basin (as discussed in Chapter 6) there are no *major* conflicts in the Tagus River basin (Llamas, 1997). However, a number of issues are worth highlighting (see also Campbell *et al.,* 2009).

The Tagus-Segura water transfer represents a complex policy of water distribution that generates many political conflicts between different regions in Spain (Ministry of Environment, 2010). Compensation measures were implemented for the reduction of stream flow and water resources availability in the Tagus River. These measures included the ending of irrigation areas under transformation and their corresponding infrastructures in the Alagón, Almonte, Tiétar and Jerte rivers, as well as the encouragement of local irrigated areas and the construction of a plant for water purification in the city of Cáceres. However, some of the plans forecasted by that time are still waiting. Within the Actuaciones para la Gestion y la Utilizacion del Agua (A.G.U.A) Programme context, water for southeast Spain is supplied by desalinisation plants (Ministry of Environment, 2010a).

An urban model based on a high consumption of resources (i.e., water resources) has led to problems of water supply in Madrid. Also, new demands such as for leisure and cleaning appear. The increasing urbanisation provokes conflicts for water distribution (i.e., irrigation vs. urban consumption). A low water quality has been observed in the river after it crosses Madrid.

**Figure 5.3** Bridge over the Tagus, north of Lisbon, Portugal (Photograph: G.D. Gooch).

Nuclear plants need a minimum base-flow to ensure their refrigeration and sometimes this minimum base-flow is not reached. Besides, such cooling measures leads to an elevation of temperature in the river that can lead to negative ecological impacts. Finally, leaks at some old power stations have been documented.

Another conflict is based upon the management of the Alcántara reservoir, just at the Spanish border with Portugal, and can be particularly critical within the transboundary context, leading to water quantity and availability issues.

## 5.5 THE INSTITUTIONAL FRAMEWORK FOR IWRM IN THE TAGUS BASIN

The occurrence of water use conflicts and the changing scenario during the last decades (climate and land-use/plant cover changes, see Chapter 7) has led to an improved shared protocol of discharges and uses of water between Spain and Portugal (Maia, 2000). The Convention on Cooperation for Portuguese-Spanish River Basins Protection and Sustainable Use (Albufeira Convention, 1998), approved by both countries, defines the framework for bilateral cooperation in topics such as freshwater and groundwater protection and the mutual co-responsibility for the sustainable use and management of the shared water resources. While the legal aspects of Albufeira Convention and IWRM within the Tagus River basin will be dealt with in the next chapter (Chapter 6), this section will examine the institutional framework that has evolved in order to managing this transboundary resource.

### 5.5.1 Transboundary arrangements

Coordination between Spain and Portugal on the Tagus takes place through the bilateral Commission for the Application and Development of the Albufeira Convention, established pursuant to Article 20 of the Albufeira Convention (CADC, 2010). The Commission is composed of delegations appointed by each party – sub-commissions and working groups may also be established were necessary. It meets in regular session once a year and decisions must be adopted by mutual consent (Art. 21). As regards duties, the Commission is, among other things, generally responsible for acting as a forum for the exchange of information; for communication, early warning and emergency systems. Aside from conducting joint studies, it also provides a forum for the assessment and consultation between the parties relating to transborder impacts and is the responsible body for adopting appropriate water quality and flow measures (Albufeira Convention, 1998). The Commission has no compliance or enforcement tools at its disposal; nor is there any mechanism for the participation of stakeholders from either nation – a permanent secretariat appears to be nascent.

**Figure 5.4** Stretch of the Tagus in Portugal (Photograph: G.D. Gooch).

The Commission defines the necessary flow regime with a view to achieving good water status, as defined in the EC Water Framework Directive (WFD, 2000; Maia, 2008), and in order to anticipate current and foreseen future uses. On February 19th 2008, during the Second Partners Conference (*Conferência das Partes*) a new hydrological regime was created between the two countries (CADC, 2010). Weekly flow measurements, with the obligation of minimum weekly flows for some of the shared rivers were agreed. For Portugal, it is evident that a more intensive use of water resources in

Spain has led to a decrease in the mean flows, an increased irregularity in the flow regime (especially with longer and more frequent dry periods), a degradation in water quality and changes in sediment transport. Overall, the Tagus River faces a number of complex strategic problems, given the opposing concerns between water and landscape conservation and increasing water demands and conflicts.

## 5.5.2 Institutional arrangements at the national level

The systems in Spain and Portugal are in some ways similar to each other. Each has a basin plan in place, and each country has a separate basin authority responsible for the management of the water resources in the basin (in Portugal, the Tejo Hydrographic Region Administration (*Administração de Região Hidrográfica do Tejo*, 2010), and in Spain, the Hydrographic Confederation for the Tajo (*Confederación Hidrográfica del Tajo*, 2010). In addition, both countries have established basin plans for their respective sections of the catchment, prepared in a collaborative and consensual way through the Tejo Basin Council in Portugal and the Tajo Council of Users in Spain. These bodies work within the general national strategies prepared by their respective National Water Councils, which are representative of national as opposed to basin interests. In Portugal, this body serves as an advisory board to the government. In Spain, it appears that the National Water Council composition is strongly dominated by representatives of the public administration, who have full control of the agenda of the meetings and the decisions taken in the forum. In Portugal, the National Water Institute (*Instituto da Água*, 2010); is responsible for the implementation of the basin plan, with the basin authority taking this function in Spain. The National Water Institute in Portugal is responsible for water management, therefore serving a similar function at the ministerial level. In both countries, there is no specific coordinating body. However, the constitution of the National Water Councils in both countries goes some way to alleviating this potential difficulty – both are made up of representatives from all relevant ministries, sectoral users, NGOs, regional authorities and technical bodies.

### *5.5.2.1 Spain*

In Spain, water resources management is controlled by the Water Authority, under the auspices of the Ministry of the Environment, and the Rural and Maritime Environment (*Ministerio de Medio Ambiente, Rural y Marino*). The Water Framework Directive requires a new hydrological plan for the Tagus basin, which requires greater participation of the regions and also more public participation (WFD, 2000). By doing this, the plan will consider all stakeholders, which amongst others include the National Water Council, and representatives from municipalities and the sectors of energy, agriculture, business associations, consumer and fishing associations and NGOs (Campbell *et al.*, 2009). In Spain the participation in developing the river basin plan under the WFD is supported by the A.G.U.A (Ministry of Environment, 2010a). In April 2008, meetings were held mainly in Madrid for different groups including: water managers at various levels of government, the Environmental Advisory Council, the National Water Council, universities and research institutes (CEDEX, IGME, CSIC), as well as private companies/industry and associations representing the interests of consumers, environmentalists, agriculture, fishing, sports, business and trade unions amongst others (Campbell *et al.*, 2009). Information was provided to stakeholders through a website as well as through brochures and information offices in Aranjuez, Cáceres, Guadalajara, Plasencia, Talavera and Toledo. Participants were informed about each step of the process and the relevant activities. For instance, several documents related to the initial steps of the planning process were uploaded with free access. The A.G.U.A Programme promotes public/social participation in the elaboration of hydrological plans (Ministry of the Environment, 2010a).

Water management in the Tagus in Spain is highly centralized in the CHT and traditionally, the water confederations have focused their activity mainly on managing the water allocation between the various users. Their main functions are: implementation of the basin Hydrological Plan; management and control of the Hydraulic Public Domain; management and control of water uses affecting two or more regions; to manage hydraulic projects such as dams and channels; functions defined by agreements with local corporations or other public and private organizations (Campbell, *et al.*, 2009). Apart from the CHT, the regional (Community of Madrid, Community of Castilla La Mancha, Community of Extremadura) and local municipal authorities also play a role in water management, as they are responsible for waste water treatment plans. Regional governments have similar responsibilities to the CHT. There are other institutions such as the 'Canal de Isabel II' (CI-II) foundation which is responsible for the water supply to the Madrid region, and management of a system of watersheds and water diversions in the northeast of the Tagus basin.

Whilst water management in the Tagus basin is highly centralized, and there was little public involvement in the past there has nonetheless been an effort for promoting transparency and public participation. Apart from what is mentioned in the basin management plan, there are a number of initiatives leaded by the CHT, the public administration and other organizations, including NGOs and farming organizations devoted to capacity building, which focus mostly on good water use practices. Some examples are campaigns promoting good practices such as those of the foundation agua-dulce.org, or the campaign for responsible consumption by the CHT.

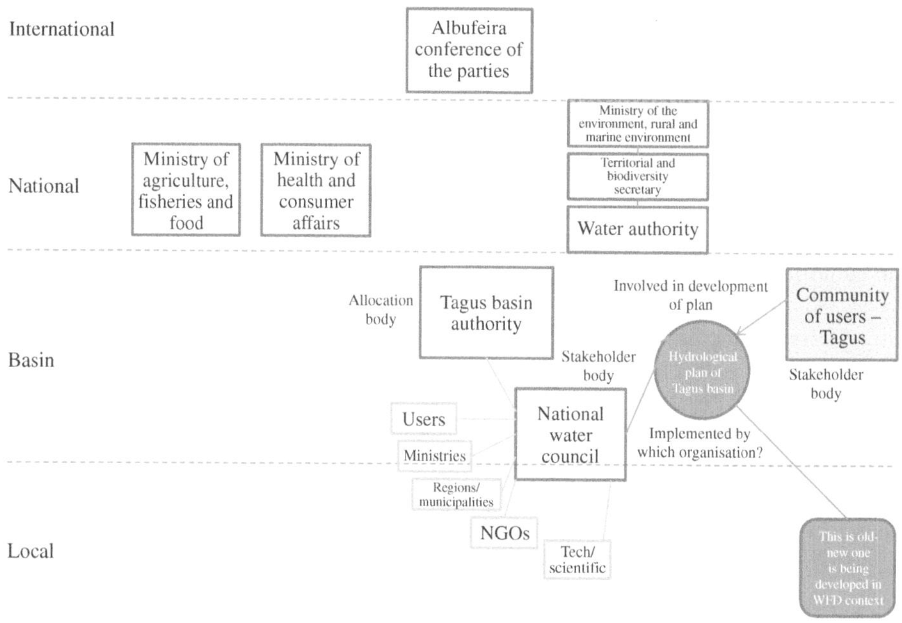


**Figure 5.5** Institutional framework for the Tagus River basin in Spain (Manasi, *et al.,* 2010).

### *5.5.2.2 Portugal*

In Portugal, the public administration institutions with competency in all issues related with water resources are the INAG, at the national level, and the ARH at the hydrographic regional level. The representation of the activity sectors and of the water users is ensured by the following two consulting or advisory entities: The National Water Council (*Conselho Nacional da Água*), CNA, and the Councils of the Hydrographic Regions (*Conselhos de Região Hidrográfica*), CRH. The former is the government forum for water resources and the latter the fora for the watersheds that belong to each one of those regions. Those who work in the latter institutions must have a background compatible with the activities that they develop which ensures the required capacity building.

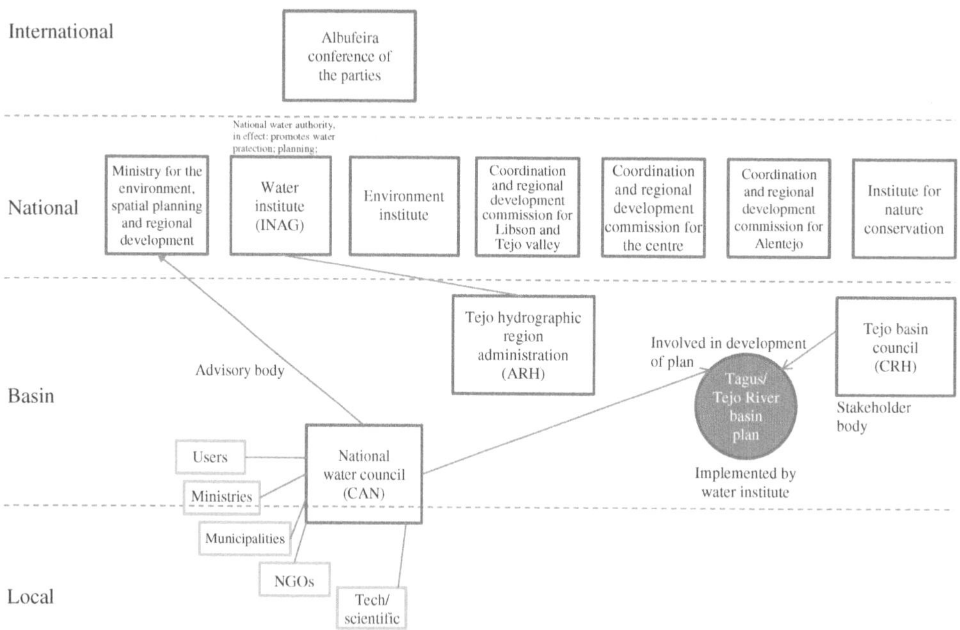


**Figure 5.6** Institutional framework for the Tagus River basin in Portugal (Manasi, *et al.,* 2010).

Also, whenever special situations occur, as during the drought of 2005, warnings are spread to the public through the media on how to conserve water or energy. Instituto da Água also produces alerts and small brochures available on the internet with some actions for specific situations, as droughts, floods and pollution incidents. They also have developed campaigns with themes related to water conservation and sustainable practices for different target audiences for instance youth or urban populations. The *Sistema Nacional de Informação de Recursos Hídricos* (SNIRH) – under the Water Institute has developed a Portuguese hydrological database – and also mounts campaigns providing information on water resources (Instituto da Água, 2010). The institute provides a Hydro Library (*Hidro Biblioteca*) which includes a large number of articles, books and publications. Civil society organizations also contribute to information and capacity building, such as agricultural or fishing organizations and NGOs involved in nature conservation such as the *Liga de Protecção da Natureza*, LPN (Nature Protection Association) and *Quercus – Associação Nacional de Conservação da Natureza* (Quercus – National Association for Nature Conservation). Information focused on environmental questions is also provided by the webpage of the Portuguese Environment Agency (*Agência Portuguesa do Ambiente*) (Instituto da Água, 2010). There is a joint program – between the Education Ministry and the Environment, Spatial Planning and Regional Development Ministry (which belongs to the Portuguese Environment Agency) – named the Eco-School Program (*Programa Eco-Escolas*) which provides environmental information to undergraduate schools and high schools, including on water resources. Eco-Centers (Eco-Centros) will be established (a few have currently been established) to disseminate environmental information from either by the Portuguese Environment Agency or by local municipalities (Instituto da Água, 2010).

### 5.5.3 Stakeholder participation in the Tagus River basin

Ensuring effective participation of relevant stakeholders within the management of the Tagus River basin constitutes a major challenge to implementing IWRM throughout the basin (Barreira, 2003). A UNECE study found that involvement of the public in environmental impact assessments for major projects, for example, was lower than was proportionate for the importance of the project (UNECE, 2005). This was not thought to be a result of lack of effort on the part of the government, but rather with the national attitude and lack of interest on the part of the populace in Spain and Portugal. Following the findings of the UNECE study, the authors encountered differing views from the experts they spoke to during the course of the STRIVER project. Broadly, the impression given by those working within the governmental/ regulatory environment – i.e. those public agencies with responsibilities for provision of environmental information – was that there was a lack of civic culture among the population, a consequence of which was that people would really only participate in decision-making where their interests were perceived to be directly at risk, especially where their property rights were under threat or where large-scale infrastructure was being considered.

At the same time, it is generally accepted that the format of the information was critical if it was to be comprehensively understood by the public. Those at the receiving end of the information provision were perhaps more generally of the view that civil servants, while improving in their ability to communicate information, did not operate within a culture of openness and that this hampered transparency overall. One view was that the Tagus context was among the worst in Spain for this because of the political complications caused by the Tagus – Segura transfer, especially when this inter-basin transfer had in effect been excluded from discussion about the pressing water use issues on the river as part of the river basin planning process for the Water Framework Directive's implementation. More encouragingly, however, the view of virtually all of those interviewed by the authors was that information provision had improved and was continuing to do so.

## 5.6 KEY CHALLENGES AND RECOMMENDATIONS FOR STRENGTHENING IWRM

Evidence from secondary sources and experts strongly suggests that the situations in Spain and Portugal have improved greatly in recent times and that they continue to improve as decision makers, stakeholders and citizens are brought closer together. Basin management has been in place in Spain for some time, with Portugal's hydrographic authorities closer to final establishment. The coordination between the two countries is stronger than ever, thanks largely to a combination of the transposition of regional legislation (which will be discussed in the next chapter), and through the closer political ties that have been established through these agreements. Problems however, remain.

This section sets out some of the key challenges and recommendations for addressing them. The recommendations are based on numerous interactions with key stakeholders within the basin. The key channels for information exchange and feedback were the stakeholder workshops organised in Toledo, 14th December 2006, Lisbon, 28th February 2008, and 16th April 2009. These workshops, and particularly the final meeting in Madrid, were instrumental in shaping the research findings into policy recommendations.

### 5.6.1 Water availability and allocation issues

As noted above there are competing water uses among different sectors within the Tagus, for example, farming, industry, urban water supply, and amongst regions, i.e., water transfers. These competing water uses place considerable demands on water availability, that in certain periods lead to water scarcity problems. Increased urbanization, land use and climate change heighten the potential for conflict between the various water uses noted about, this is particular evident in during water shortage periods.

**Figure 5.7** On the Tagus River, Portugal (Photograph: G.D. Gooch).

Related to this issue is the finding that in Spain, water saving upstream does not necessarily affect water availability downstream because of the Júcar and Segura basin transfers. Basin transfers pose a significant challenge for the Tagus River basin. In Spain, the Tagus-Segura water transfer accounts for 10 percent of the total water use in the Spanish part of the Tagus River basin, averaging 500 $hm^3$ per year. A key question is therefore whether the current rate of transfer can be met whilst also ensuring for the long-term viability of the resource.

Urban demands for water use are a key challenge in both Spain and Portugal. Lack of water availability in Madrid in the future is a real possibility, and future pressure on the headwaters of the Tagus due to urban demands in primarily for Lisbon's population is considerable.

One issue where significant efficiency potential might occur relates to non-controlled water abstraction, which is causing over-exploitation in some areas.

Water efficiency savings can be made through a number of strategies. All water consumptions must be measured and not just estimated. The consumer pays principle must be fairly implemented, and should take into consideration the social, environmental and economic consequences. Water pricing should be designed to promote efficient use of water resources. Distortion due to intra-sectoral competition between the two countries should be prevented. More incentives for water saving and reuse should be in place, and new regulations related to 'hydro-efficiency' similar to schemes for energy efficiency could provide a significant contribution.

A further issue relates to how water allocation values are determined and managed. Water allocation has traditionally been based mainly on long-term average values and assumes normal management conditions. Adaptation of the existing water systems to the concrete circumstances of each year is done reactively rather than adopting an anticipatory approach. Those systems may therefore fail to accommodate drought periods effectively. However, Spain and Portugal have reached an agreement recently to change water allocation values from fixed annual volume to seasonable flows.

Basin plans are being developed throughout the EC in accordance with the WFD. The existing basin plans do not fully account for the effects of climate and land cover changes on water availability. The next planning cycle must therefore take account of water quantity and water quality collectively, given that the quality determines the availability of water use for certain the various uses. River basin plans must also account for the effects of climate and land cover change.

A crucial water management issues that must be addressed in the Tagus is the maintenance and improvement of hydrological monitoring systems, which is a prerequisite for ensuring that IWRM is implemented effectively. More resources must be put in place in order to promote the maintenance and improvement of the monitoring systems.

### 5.6.2 Water quality and ecological status

Many problems threaten water quality within the Tagus basin, such as deficient water treatment, especially in small towns; over-exploitation, which increases the concentration of pollutants and nutrients; and diffuse pollution sources. In Madrid and the metropolitan Lisbon areas (including Lisbon, Barreiro, Almada and Montijo), for example, lack of capacity to treat all wastewater is a major challenge, even to accomplish the goals of EC Directives. The lack of capacity to treat wastewater also contributes to significant pollution of the estuary although the construction of a new treatment plant is about to be completed.

**Figure 5.8** STRIVER Tagus, Stakeholder Workshops, Lisbon (Photograph: G.D. Gooch).

There are clear linkages between qualitative and quantitative issues. To date enforcement of illegal water extraction has been weak and should therefore be strengthened. There is also a clear need to increase investment in wastewater treatment, and control wastewater dumping, gravel excavations and effluents from farms and irrigations systems.

A more tailored system for water quality management would improve matters. Water quality should therefore be adapted to use requirements, for example, industrial uses might be able to tolerate lower water quality than drinking water. Problems with alien invasive species are also evident in various parts of the Tagus.

A number of water quality issues could be tackled by a strict implementation of the requirements of the Water Framework Directive. Clearly, ecological quality evaluation must be implemented in order to verify if the good status objectives of the EC Water Framework Directive will be met by 2015. An ecological flow regime should be implemented in order to achieve good ecological status/ potential. The polluter pays principle must also be implemented effectively and fairly. Within this regard, lessons could be learnt from air pollution regimes.

### 5.6.3 Stakeholder participation in water management

Although a significant effort has been made in the last few years to make all technical data available to the public, information about the decision-making process is still deficient. Transparency in decision-making can help ensure a balanced use of water and reduce negative social and environmental impacts. More effort is therefore needed to facilitate and promote stakeholder involvement in decision-making. Management at local levels needs to be coordinated effectively to fully account for public trust.

Capacity building is a pre-requisite to public participation. However, more need to be done to ensure that basic education on water issues is provided. Water users should be provided with more education campaigns focused on promoting best practices. At the same time, education and information tools should be made available (e.g., teaching materials, 'water footprint calculator', etc.) Real time information on the quality of drinking water could be implemented in order to make people conscious of water quality and motivate public trust.

Water resources management has been considered during decades a predominantly technical issue, so public participation was not incorporated in the past, although notable developments have occurred in recent years through the adoption of the Water Framework Directive. However, the question of *who* participates in *what* decisions and *how*, remains a key challenge. Active involvement and public consultation in decision-making remains largely an aspiration rather than an effective tool by which to manage water resources equitably and sustainably.

### 5.6.4 Legal and institutional aspects

Major efforts have been made to put a legal framework in place that facilitates good water governance and IWRM – although notable challenges remain. There is a need for stronger support to monitor implementation of laws relating to access to information, participation and IWRM. In addition water basin authorities should have the necessary resources to monitor and enforce regulations, as ineffective law enforcement constitutes a major barrier to the implementation of IWRM within the basin. Capacity building, stronger institutional coordination and awareness raising is also needed to improve the implementation of IWRM law and policy.

Overlapping competences between different authorities, as well as a lack of dialogue and coordination of policies can lead to inefficiency and inequity since weak institutions can result in the allocation of water being determined largely on the basis of power. In Spain, different interests are represented by the Ministry of the Environment and Rural and Marine Affairs. In Portugal, intra-sectoral coordination is needed between the different ministries directly or indirectly involved.

Intra-river basin management and coordination between Spain and Portugal is at an embryonic stage. Co-ordination of the Spanish and Portuguese Tagus River basin authorities should be promoted further, and transparency to the public improved. Supranational coordination, considering the entire Tagus basin, should be enforced in the elaboration of each of the new River Basin Management Plans. The production of a final joint document is desirable. Long term policies independent of government shifts are needed. Policy-makers must consider the long-term ecosystem impacts and the impact on flows in Portugal of water transfers to the southeast of Spain. In addition, the interstate Albufeira Convention does not sufficiently account for water quality issues, or management of extreme events, the Convention's provisions relating to water quality and the co-ordinated management of extreme events should therefore be strengthened.

## 5.7 REFERENCES

Albufeira Convention (1998) Agreement on Cooperation for the Protection and Sustainable Use of the Waters of the Spanish-Portuguese Hydrographic Basin, Nov. 30, 1998, (entered into force Jan. 17, 2000), 2099 UNTS 275.

Administração de Região Hidrográfica do Tejo I.P. (2010) http://www.arhtejo.pt/index.html

Beguería, S., Raju, K.V., Manasi, S., Latha, N., Umesh Babu, M.S. and Nagothu, U.S. (2008) Land Use and Land Use Change – Implications for Water Resources and Water Use in the Tagus and Tungabhadra Basins. STRIVER Task Report No. 9.3 , www.kvina.niva. no/striver/Disseminationofresults/STRIVERReports/tabid/80/Default.aspx

Barreira, A. (2003) The Participatory Regime of Water Governance in the Iberian Peninsula. *Water International* **28**(3), 350–357.

CADC (2010) Commission for the Application and Development of the Agreement on Cooperation for the Protection and Sustainable Use of the Waters of the Spanish-Portuguese Hydrographic Basin, http://www.cadc-albufeira.org/pt/comision.html.

Campbell, D., Rieu-Clarke, A., Allan, A., Gooch, G.D., Stålnacke, P. and Nagothu, U.S. (2009) Final Stakeholder Analysis – Stakeholder Participation in the STRIVER Basins. STRIVER Report D4.2. 81p, http://kvina.niva.no/striver/Disseminationofresults/tabid/70/Default.aspx

Confederación Hidrográfica del Tajo (2010) http://www.chtajo.es

EC (1979) Council Directive 79/409/EEC, of 2nd April 1979 on the conservation of wild birds, OJ L 103, 25 April 1979, pp. 1–18.

Fernandes, T.F., Elliott, M. and Da Silva, M.D. (2005) The Management of European Estuaries: A comparison of the features, controls and management framework of the Tagus (Portugal) and Humber (England). *Aquatic Ecology* **29**(3–4), 459–468.

Instituto da Água (2010) http://www.inag.pt

Llamas, R. (1997) Transboundary Water Resources in the Iberian Peninsula. In *Conflict and the Environment*. N.P. Gleditsh (ed.), 335–354, Kluwer Academic Publishers, The Hague.

Maia, R. (2000) Portuguese-Spanish River Basins: Bilateral Agreements' evolution and context. *Water Science & Technology* **42**(1), 227–233.

Maia, R. (2008) The EU Waterframework Directive implementation in the Iberian context. Paper presented at 13th IWRA World Water Congress, 1–4 September 2008, Montpellier, France.

Manasi, S., Joy, K.J., Paranjape, S., Nagothu, U.S., Campbell, D., Latha, N., Portela, M.M., Betamio de Almeida, A., Machado, M, Raju, K.J. and Beguería, S. (2010) The Science-Policy-Stakeholder Interface (SPSI) in Land- and Water-Use Interactions. In *Science, Policy and Stakeholders in Water Management – An Integrated Approach to River Basin Management*. G.D. Gooch and P. Stånacke (ed.). Earthscan, London.

Confederación Hidrológica del Tajo (2001) Plan Hidrológico de la cuenca del Tajo, Confederación Hidrográfica del Tajo (CHT), Madrid.

Ministry of the Environment, Rural and Marine Environment (2010) Water Resources – Tagus-Segura Transfer, http://www.chsegura.es/chs_en/cuenca/resumendedatosbasicos/recursoshidricos/trasvaseTajoSegura.html

Ministry of the Environment, Rural and Marine Environment (2010a) Actuaciones para la Gestión y la Utilización del Agua, http://www.mma.es/secciones/agua/entrada.htm

Nesheim, I., McNeill, D., Stålnacke, P., Sekhar, N.U., Grizzetti, B., Allen, A.A., Barton, D., Beguería-Portugés, S., Berge, D., Bouraoui, F., Campbell, D., Deelstra, J., García-Ruiz, J.M., Gooch, G.D., Joy K., Lana-Renault, N., Machado, M., Manasi, S., Nhung, D.K., Paranjape, S., Portela, M.M., Rieu-Clarke, A., Saravanan, V.S., Thaulow, H. and Vicente-Serrano, S. (2008) *The first IWRM assessment report for the four case basins: Glomma, Tagus, Sesan and Tungabhadra.* STRIVER Report D5.1.

Ramsar Convention (1971) Convention on Wetlands of International Importance especially as Waterfowl Habitat, 2 Feb. 1971 (entered into force 21 Dec. 1975), 996 UNTS 245.
UNECE (2005) Meeting of the Parties to the Convention on Access to Information, Public Participation in Decision-Making and Access to Justice in Environmental Matters – Implementation Report (Portugal), UN ECE Doc. ECE/MP.PP/2005/18/Add.18, 27 Apr. 2005, http://www.unece.org/env/documents/2005/pp/ece/ece.mp.pp.2005.18.add.18.e.pdf
UNEP (2004) Freshwater in Europe – Facts, Figures and Maps, UNEP/DEWA, Châtelaine.
WFD (2000) Directive 2000/60/EC of the European Parliament and of the Council establishing a framework for Community action in the field of water policy, Official Journal L 327, 22/12/2000 pp. 0001–0073.

# Chapter 6

# The legal framework for the Tagus River basin

A. Rieu-Clarke and A. Allan

## 6.1 INTRODUCTION

The legal framework provides an essential platform for the effective management of waters, particularly within a transboundary setting. The purpose of this chapter is therefore to examine the extent to which key principles of IWRM and good governance have been incorporated into the legal framework applicable to the Tagus River basin. The chapter will therefore look at the range of legal instruments relating to the management of transboundary waters that have been adopted by both Spain and Portugal. Not only will such an analysis account for the national legislation that is in place, but regard will also be paid to the international level, including bilateral agreements concluded between the countries, and international and regional agreements adopted under the auspices of the UN, the UN Economic Commission for Europe (UN ECE), the European Community and others.

The chapter finds that Spain and Portugal have a long history of developing laws related to transboundary water resources management, and more recently a major influence has come from the regional level in the shape of UN ECE and EC laws. The adoption of the recently amended 1998 Albufeira Convention aimed at the protection of the surface waters and groundwater and the aquatic and terrestrial ecosystems that directly depend on them and for the sustainable use of the water resources of the hydrographic basins of the river Miño, Limia, Douro, Tagus and Guadiana constitutes a major milestone in the development of the legal framework within the basin (Spain-Portugal Agreement, 1998). In addition, the national legislation, underpinned by regional laws (EC and UN ECE) has provided a solid basis to ensure that good governance principles of transparency, participation and accountability are supported. Ultimately, the chapter concludes that the necessary law is in place; what is now needed is support to ensure that the 'spirit of the law' is fully realised.

## 6.2 INTERNATIONAL, REGIONAL AND BI-LATERAL COMMITMENTS

### 6.2.1 Multilateral agreements

Spain and Portugal are both parties to a range of multilateral environmental agreements that relate to the management of water resources. The most notable global multilateral agreement in relation to transboundary watercourses, to which both Spain and Portugal are party, is the 1997 UN Watercourses Convention (Watercourses Convention, 1997). This applies to the uses, protection, preservation and management of international watercourses and their waters. Under the substantive rules and principles of the Convention, states must utilise and protect their international watercourses in an equitable and reasonable manner, taking into account all relevant factors and circumstances, with no one use enjoying inherent priority over other uses. States are also under an obligation to take all appropriate measures to prevent the causing of significant harm to other watercourse States. The Convention further provides that States must protect and

©2010 IWA Publishing. *Integrating Water Resources Management: Interdisciplinary Methodologies and Strategies in Practice*. Edited by Geoffrey D. Gooch, Alistair Rieu-Clarke and Per Stålnacke. ISBN: 9781843393252. Published by IWA Publishing, London, UK.

preserve ecosystems of international watercourses and the marine environment, as well as prevent, reduce and control pollution and the introduction of alien species. From a procedural standpoint, the 1997 UN Watercourses Convention requires States to regularly exchange data and information on the condition of the watercourse, and notify potentially effected States of planned measures that may be inconsistent with the principle of equitable and reasonable utilisation. In addition, State parties are obliged to enter into consultations where a planned measure may be inequitable or unreasonable, and resolve their disputes in a peaceful manner. While the Convention has not yet entered into force, many of the rules and principles contained therein are considered to represent existing customary international law.

Other relevant global environmental agreements to which both Spain and Portugal are party to include the 1971 Ramsar Convention (Ramsar Convention, 1997), the 1992 Biodiversity Convention (Biodiversity Convention, 1992), the 1992 Climate Change Convention (Climate Change Convention, 1992) and the 1994 Desertification Convention (Desertification Convention, 1994).

The Tagus Estuary is designated a wetland of international importance under the Ramsar Convention (Ramsar Convention, 1997). Pursuant to the provisions of the Convention contracting parties are obligated to promote the wise use of wetlands within their territory, and promote the conservation of wetlands listed as of international importance under the Convention. "Wise use" is defined under the agreements as being, "sustainable utilisation for the benefit of mankind in a way compatible with the maintenance of natural properties of the ecosystem." In relation to transboundary waters, contracting parties are obliged to consult with each other concerning the implementation of obligations arising from the Convention especially in the case where a water system is shared by contracting parties. States must also endeavour to coordinate and support present and future policies and regulations concerning the conservation of wetlands and their flora and fauna.

Under the Biodiversity Convention (1992), states have a duty to integrate the conservation and sustainable use of biological diversity into relevant plans and policies, to identify processes and activities that are likely to have a significant adverse impact on conservation and sustainable use of biodiversity; and to ensure that environmental impact assessment of proposed projects and activities are conducted. Where activities in one State jurisdiction are likely to have an impact in another State, contracting parties must promote notification, exchange of information and consultation measures.

Pursuant to the Climate Change Convention (1992) states must develop and elaborate appropriate and integrated plans for coastal zone management, water resources and agriculture, and for the protection and rehabilitation of areas affected by drought and desertification, as well as floods. Similarly, the Desertification Convention (1994) seeks to combat desertification and mitigate the effects of drought in countries experiencing serious drought and/ or desertification. States agree to develop long-term integrated strategies that focus simultaneously, in affected areas, on improved productivity of land, and the rehabilitation, conservation and sustainable management of land and water resources.

## 6.2.2 Regional agreements – the UN ECE

Under the UN ECE Treaty Regime Spain and Portugal are under a range of obligations that relate to transboundary water resources management and under the UN ECE Helsinki Convention (1992), both Spain and Portugal are obliged to take all appropriate measures to prevent, control and reduce transboundary impact. "Transboundary impact" is defined in the Convention as, "any significant adverse effect on the environment resulting from a change in the conditions of transboundary waters caused by a human activity, the physical origin of which is situated wholly or in part within an area under the jurisdiction of a Party, within an area under the jurisdiction of another Party. Appropriate measures described within the text of the Convention, include the establishment of permit and licensing schemes for waste water discharges, the application of best available technology to reduce point source pollution, and the implementation of best environmental practices for non-point source pollution solutions. The agreement also places States under certain procedural and institutional obligations, including requiring that transboundary watercourses are covered by bilateral and multilateral agreements, and the establishment of joint bodies for the implementation of the treaty commitments.

A further notable UN ECE Convention to which both Spain and Portugal are signatories is the Aarhus Convention, which was adopted in 1998 (Aarhus Convention, 1998). The Convention obliges parties to take the necessary legislative, regulatory and proper enforcement measures to secure access to information, public participation and access to justice in environmental matters. Both Spain and Portugal are also party to the 1991 UN ECE Convention on Environmental Impact Assessment in a Transboundary Context, which obliges contracting parties to establish environmental impact assessment procedures that permit public participation, with respect to proposed activities that are likely to cause significant adverse transboundary impact (UN ECE Espoo, 1991).

### 6.2.3 Regional agreements – the EC

As members of the EC, both Spain and Portugal are also obliged to implement the provisions of the relevant EC legislation, most notably the EC Water Framework Directive (WFD, 2000). The EC WFD was adopted in 2000 with the aim of establishing a framework for the protection of inland surface water, transitional waters, coastal waters and groundwater (WFD, 2000; Rieu-Clarke, 2004). By 2015, and subject to exceptions, EC Member States should seek to achieve good water status in all EC waters. The Directive also requires Member States to identify river basin districts and assign a competent authority responsible for the implementation of the Directive within each of the districts. In addition, a river basin management plan must be developed for each of the river basin districts. The management plan should be designed to ensure that the provisions of the Directive and related EC legislation are implemented at the basin level. The Directive also encourages member states to ensure the active involvement of all interested parties in the implementation of the Directive, and in particular through the production, review and updating of river basin management plans. The WFD has played a leading role in promoting new environmental policies in both Spain and Portugal (Maia, 2008) and to ensure efficient use of water some conservation measures linked to the WFD have been developed. Both Spain and Portugal apply the user-pays approach based on costs and benefits, and Spain has incentives for the adoption of new water saving technologies and efficient transport, and of non water-intensive crops and cultivars. Nonetheless there is still much waste of water resources because of intensive agricultural practices for instance.

Spain adopted the hydrological plan of the Tagus basin in 1999 which covers the following: ensuring present and future water demand; implementing efficient water management; protecting water resources and the environment; ensuring water quality; protection from hydrological extreme events; and protection of cultural and recreation uses. Portugal has developed a river basin plan that provides a comprehensive description of the water resources availability, as well as the general framework, strategies and the programmatic targets for the future water resources planning activities.

In Portugal, River Basin Management Plans (RBMPs) have been replaced by Hydrographic Region Management Plans (HRMPs). According to the new Portuguese Water Law, the Portuguese HRMPs should provide support to the management, protection and to the environmental, social and economic valuation of water resources – including the estuaries, the coastal areas and the aquifers, as well as the surface waters. The HRMPs will become part of the municipal master plans, replacing the river basin plans from 2001/2002. The HRMPS will also fulfil reporting requirements to the European Commission in the application of the WFD, in terms of the fulfilment of the environmental objectives in 2015, and they will also provide public information.

Some of the aspects embedded in the HRMPs are: integrated assessment of technical, economic, environmental and institutional issues; the supply and demand in order to conserve water resources; coordination between regional planning and land management and the use of environmental and conservation measures; the need to involve the economic agents and the public as a way to obtain a wider consensus.

Each HRMP must include a general description of the corresponding hydrographic region; a characterization of the existing pressures, both natural and resulting from human activity; and a programme of measures envisaging the pursuit of the environmental objectives established in the new water law. According to this law, the plans should be the object of public consultation during the development, revision and updating phases. While the river basin plans developed between 2001 and 2002 were mostly use/quantity-goal oriented, the HRMPs are essentially environment/quality goal oriented and mark a new era of proactive management. As discussed in Chapter 8, according to the WFD and the new Portuguese Water Law, new public authorities were created and called upon to assist in the planning and execution of regulations and measures for the implementation of sustainable water use. Amongst these authorities the most important ones are the recently created (in October 2008) Hydrographic Region Administrations (HRAs; 10 in number), and the existing, though restructured, National Water Institute (INAG). The HRAs, like the INAG, belong to the Ministry for Environment, Spatial Planning and Regional Development and are regional entities, specializing in water issues and with administrative and financial autonomies. The HRAs are there to promote the HRMPs.

### 6.2.4 Bi-lateral agreements

Treaty practice relating to Spain and Portugal's transboundary watercourses dates back to the Limit Treaty of 1864, (concluded by an exchange of notes in 1912) which stipulated that water within the border regions of the countries had to be utilised for mutual benefit. In 1927 an Agreement to regulate hydropower on the Douro River, was concluded. The latter agreement became obsolete on the adoption of a further agreement on the development of the Douro for hydropower in 1964. A further agreement in 1968 extended the provisions of the 1964 agreement to other the international sections of the Miño, Limia, Tagus, Guadiana, and Chanza and their tributaries (Spain-Portugal Agreement, 1968).

Most recently, the 1998 Albufeira Convention was adopted by Spain and Portugal (Spain-Portugal Agreement, 1998). The 1998 Albufeira Convention is applicable to, "the hydrographic basins of the rivers Miño, Limia, Douro, Tagus and Guadiana" (Art. 3(1)). "Hydrographic basins" is defined in the agreement as meaning, "the land area from which all the surface runoff flows through a series of streams, rivers and, in some cases, lakes towards the sea, entering it through a single river mouth, an estuary or a delta, and also associated groundwater (Art. 1(1)(b))." The Albufeira Agreement provides the provisions of the 1964 and 1968 Agreement remain in force so far as they do not run contrary to the provisions of the 1988 Agreement (Art. 27(2)). The purpose of the Agreement is, "the protection of the surface waters and groundwaters and the aquatic and terrestrial ecosystems that directly depend on them and for the sustainable use of the water resources of the hydrographic basins" (Art. 2(1)).

The principle of equitable and reasonable utilisation is not specifically provided for within the text of the Albufeira Convention (Spain-Portugal Agreement, 1998). However, as noted in Art. 2(1), the purpose of the Convention is, "the protection of the surface waters and groundwaters and the aquatic and terrestrial ecosystems that directly depend on them and for the sustainable use of the water resources of the hydrographic basins." In accordance with its purpose, the Albufeira Convention recognises the right of each of the parties to the sustainable use of water resources, as well as their obligation to protect such waters, and implement measures to prevent, eliminate, mitigate and control transborder impacts (Art. 15(1)). The Agreement also obliges the parties to define, "the flow regime necessary to guarantee the satisfactory conditions of the waters (Art. 16(1))." In addition, the parties must, "adopt, individually or jointly, the technical, legal, administrative and other measures" necessary to, *inter alia,* achieve satisfactory condition of the waters; prevent degradation of the waters and control pollution; prevent, eliminate, mitigate or control transborder impacts; ensure that the use of the water resources are sustainable; promote rational and economic use. In terms of water quality and pollution, the Agreement also obliges the parties to comply with quality targets and standards of waters classified under EC law. The substantive provisions of the Convention are very much in line, and clearly influenced by, the regional agreements noted above, as well as customary international law.

As regards the exchange of information between riparian states, the Parties to the 1998 Albufeira Convention, through the Convention's Commission, must regularly and systematically exchange data and information on matters covered by the Convention, including the management of the waters of the basins; activities that may have a transboundary impact; and legislation, organisational structures and administrative practices. Detailed provisions are contained in Annex 1 (Spain-Portugal Agreement, 1998). The parties are also obligated to provide the Commission with, "all the information it needs to fulfil its terms of reference and responsibilities", including information on how the activities of the Convention are implemented nationally, and those activities that might have a transboundary impact. In addition, the parties are obliged to produce an annual report on relevant activities and an update on the status of national implementation. With respect to extreme events, the Albufeira Convention also obliges Spain and Portugal to, "establish or improve joint or coordinated communication systems to transmit early warning or emergency information, to prevent or correct the situation in question and to take pertinent decisions." The parties must provide information on the entities and procedures within each of their countries for the transmission of information on early warning and emergency situations. With respect to public rights of access to information under Albufeira, the parties are obligated to "create the conditions to make available to anyone who submits a reasonable applicable the requested information on matters covered by" the Convention, but may refuse to do so if the following will be affected: national security; the confidentiality of procedures carried out by public authorities; the international relations of the State; the general security of the population; the confidentiality of legal proceedings; or commercial / industrial confidentiality (Art. 6).

The most recent version of the Convention is currently in the ratification process, and includes greatly increased details on the flow regimes applicable to each of the basins it covers, including daily flow rates where appropriate. Common cartography and modelling approaches have been adopted to further enhance the monitoring of flow and quality at the border.

## 6.3 NATIONAL LAWS

Laws that related to IWRM in the Tagus at the national level cover a wide range of aspects. While the legal regime for the management of water resources *per se* will be important, other laws that relate to key principles of good governance, e.g. transparency, accountability and participation, will also influence the effective implementation of water laws. The focus of this section will therefore be to adopt an expansive definition to not only look at the applicable water laws within Spain and Portugal, but also to consider how the wider legal system supports the implementation of IWRM through the provision of a framework of good governance.

### 6.3.1 Law and policy development

As noted above both countries are bound by the terms of the Water Framework Directive, and from that perspective are therefore obliged to establish basin management administrations and to manage pollution control and abstractions at the basin level.

Basin management has existed in Spain since 1926, and this has therefore made it receptive to the transposition of the Water Framework Directive (WFD, 2000). In Spain, IWRM has been implemented largely through the amended Spanish Water Act (Royal Decree, 2001), which integrates surface and ground waters and reaffirms the catchment as the basis for water management.

In Portugal, however, detailed regulations for basin management do not traditionally exist, resulting in the poor application of the existing law and difficulties in complying with EC legislation. For instance, Portugal was one of five member states in a case before the European Court of Justice (ECJ) for failure to implement certain provisions of the Water Framework Directive (ECJ, 2006); see also case C-251/03 where the ECJ condemned Portugal for not complying with the quality parameters laid down in the Drinking Water Directive (ECJ, 2005). However, a major step forward has been the recent adoption of the Water Framework Law in 2006, which has transposed the commitments of the WFD into Portuguese national legislation (Decree, 2006).

### 6.3.2 Water allocation and prioritization measures

In Portugal, the water planning instruments are designed for *inter alia* defining criteria for "the prioritisation of the various water uses taking into account the economic value of each of them". To mitigate against the impacts of droughts, the Portuguese Water Framework Law establishes that priority shall be given to domestic uses, and to the vital sectors of agriculture, livestock and industry (Decree, 2006). In addition, Decree 226-A/2007 (Decree, 2007) defines the legal regime of water uses including the granting conditions. In Spain, this prioritisation is set out in Art. 60 of the Water Law (Decree, 2006), namely:

(1) domestic supply, including water for industries of low water consumption located in the population centres and connected to the municipal network,
(2) Irrigation and other agricultural uses,
(3) Industrial uses for electricity generation,
(4) Other industrial uses not mentioned in the previous numbers,
(5) Aquiculture,
(6) Recreational uses,
(7) Navigation and aquatic transport, and
(8) Other uses.

Problems have been identified with respect to the process in place for the allocation of water use rights in Spain, however, with applications for uses taking a significant period for completion. The restrictions on changing water use, even with respect to changes of crops, are very tight and are potentially inimical to the variation and flexibility that will be needed in the context of climate change and other forces affecting the availability of water.

### 6.3.3 Protection of aquatic ecosystems

In Portugal, the Water Framework Law provides that the protection and restoration of aquatic ecosystems is of major importance (Decree, 2006). The law stipulates that there must be proper planning and regulation of the water uses in order to protect the aquatic ecosystem and the law contains specific measures in that regard.

The protection and restoration of aquatic ecosystems is also embedded in Spanish law. In line with the WFD, Art. 92 therefore stipulates that one of its objectives is to protect aquatic ecosystems through the prevention of further deterioration in the condition of water bodies. In addition, the Article requires that, where necessary, water bodies are improved, with a view to the further protection of aquatic ecosystems, terrestrial ecosystems and wetlands. Art. 40 of the Water Law supplements the latter obligation by providing that hydrologic plans must, rationalise the use of water in harmony with the environment and other natural resources (Royal Decree, 2001).

### 6.3.4 Broader governance arrangements that influence IWRM law

As noted above, both Spanish and Portuguese law have gone a long way to incorporate IWRM principals through the transposition of the WFD into national legislation. However, a crucial component in the successful implementation of the commitments contained in the WFD is to ensure that good governance principles are supported at the national level. The focus of this section is therefore to consider whether such principles are supported within national law. Given the legal focus of the chapter, it will fall short of assessing whether the necessary social, economic and political conditions are in place to guarantee that the legal provisions are fully complied with.

As noted above, both Spain (29th Dec. 2004) and Portugal (9th June 2003) have ratified the Aarhus Convention (Aarhus Convention, 1998). Both countries are also obliged under EC law to transpose the terms of the Aarhus Convention at the national level (Directive, 2003; Directive, 2003a). Former UN Secretary-General Kofi Annan, has described the Aarhus Convention and the principles it embodies, as 'the most ambitious venture in the area of "environmental diplomacy" so far taken under the auspices of the United Nations (Stec and Casey-Lefkowitz, 2000). Similarly Stec and Casey-Lefkowitz (2000), describe the Convention as 'a global framework for strengthening citizens' environmental rights'. The section will therefore focus on the three pillars of good governance articulated in the Convention, namely access to information, participation in decision-making and access to justice (see Rieu-Clarke *et al.*, 2010). These three pillars of the Convention access to information, participation in decision-making and access to justice, will be considered individually within the context of Portuguese and Spanish legislation.

### 6.3.4.1 Access to information

The Aarhus Convention obliges public authorities, in response to a request for environmental information, "to make such information available to the public, within the framework of national legislation", without an interest having to be stated (Art. 4(1)). Such information must generally be made available within one month after the request has been submitted. Certain rights of refusal are included within the Aarhus Convention, but these must be interpreted in a restrictive manner.

A general right to information (and a right to be informed without discrimination) is provided for in Portuguese law, without an interest having to be stated. A similar position is taken in Spain. In addition, Portugal must ensure that environmental information is disseminated, with access being actively promoted. Such an active approach to dissemination of information, rather than simply assisting with requests for information, does not appear to have been adopted in Spain, which has opted for a more restricted response-based approach.

The grounds for refusing a request for environmental information mirror those required under the Aarhus Convention (Aarhus Convention, 1998). Confidential information is defined in Portugal in Act 67/98 (Act, 1998), which includes proceedings or information held by or for public authorities, commercial and industrial information and personal data (unless consent has been given). Additionally, national law clearly defines the grounds for refusal which in accordance with Art. 11(8), Law 19/2006 "shall be interpreted in a restrictive way by the public authorities, taking into account for the particular case the public interest served by disclosure of the information and the interests served by the refusal". Requests for environmental information may be refused "when the information requested is not held by or for the public authority to which the request is addressed". Art. 11(6) more specifically provides that the request for information may be refused if disclosure of the information would adversely affect: confidentiality of proceedings, international relations, public security or national defence; ongoing investigations; confidentiality of commercial or industrial information to legally protect a legitimate economic interest, as well as the public interest in maintaining statistical confidentiality and tax secrecy, intellectual property rights, and finally protection of the environment to which such information relates, for instance, the location of rare species. The Spanish grounds for refusal include:

(1) where the information requested is not held by the authority,
(2) the request is manifestly unreasonable or formulated in too general manner,
(3) the request concerns material in the course of completion or documents or unfinished data,
(4) the request concerns internal communications, taking into account the public interest served by disclosure.

A request for environmental information may also be refused if the disclosure would, adversely affect,

(1) the confidentiality of the proceedings of public authorities, where such confidentiality is provided within a the law,
(2) international relations,
(3) national defence or public security,
(4) the course of justice,
(5) the ability of a person to receive a fair trial or,
(6) the ability of a public authority to conduct an enquiry of a criminal or disciplinary nature (Article 13.2, Law, 2006).

Spain also applies the same grounds as are included in Art.11(6) of the Law 19/2006, under Art.13(2) of Law 27/2006, and applies a similar obligation to interpret grounds for refusal restrictively.

Portugal requires that responses to requests be made within ten days normally, but not more than one month. This rises from one month to two months in Spain, but only in periods of exceptional volume of requests. In relation to the publication of the factors being taken into account in policy making, both countries are bound by the terms of Directive 2003/35 on public participation in respect of certain plans and programmes, which sets out responsibilities for the publication of and involvement of the public in the preparation of plans and programmes which would include those relating to water (Directive, 2003). In addition, the development of river basin management plans under the WFD obliges authorities to adhere to certain standards in consultation and involvement of stakeholders.

In Portugal, while the national water authority must collect information on the performance of water-sector organisations, it is under no obligation to make this information public. It is unclear if this is also the position in Spain. In both countries, it seems that the support for information availability and accessibility are concentrated on ensuring that public authorities respond properly and effectively, rather than imposing duties on them to actively disseminate and support public involvement in the relevant decision-making processes. It also seems that there have been problems with accessing data on the internet in both Spain and Portugal, although this has improved a great deal in recent years. Delays in responding to requests for information are (or at least were in 2005) the norm rather than the exception in Portugal, despite the best intentions of the legislation. A similar position is seen in Spain, with a lack of human resource commitment being specifically raised in relation to responsible authorities.

With respect to the collection of information with relevance to IWRM, INAG is required in Portugal to collect all information on water, but no specific legal provisions exist with respect to the frequency of that collection or its provision to others. It should, however, provide an annual report on the state of the environment. In Spain, authorities are required to publish annual reports on the state of the environment, with full reports due every four years (including non-technical summaries). Questions have been raised regarding the quality of information that is available to compilers of these environmental reports, and the UNECE found that in Portugal, for example, that the monitoring network on which data was based was less than adequate for the task.

Problems with exchange of data relevant to IWRM are also evident in the law – detailed procedures for the transfer and exchange of this type of information between responsible authorities are generally absent, although mandatory systems of information transfer to public authorities do exist in the context of flooding, for instance.

#### *6.3.4.2 Participation in decision-making*

Both Portugal and Spain have protected the rights of freedom of expression and freedom of association within their Constitutions. Spanish citizens also have the right to participate in public affairs, so the basis for public participation is appropriately laid out. The constitutions of both also provide that equality before the law is secured, with discrimination on the basis of gender, among other things, prohibited. The Portuguese Constitution also states that "direct and active participation of men and women in public life is a fundamental condition and instrument of the democratic system and that the law shall promote gender quality" (Constitution, 2005), and in Spain the government is subject to the principle of equality of gender in the appointment of senior members of civil service entities. Portugal provides for 33% representation of women.

Under Portuguese Law, and in accordance with Art. 7(6) of law Decree 232/2007, plans and programme proposals that are likely to have significant effect on the environment – along with an environmental report – must be submitted to public consultation. The public consultation must have the aim of soliciting suggestions and observations from associations, organizations or non-governmental groups and others that have any interest or might be affected by the proposals. The justification for qualify a plan or programme as having a significant adverse effect, or not, must also be made available to the public. This general right of participation in environmental decision-making in Portugal is supplemented by the Water Framework Law, which foresees public participation in the elaboration review and assessment of water planning instruments (Decree, 2006).

The obligation to publish the facts and its analyses for framing major policy related to IWRM is not dealt with under Spanish Law. However, Law (2006) requires public participation in the preparation of plans, programmes and general regulations in environmental issues. Therefore the public should be informed by public notice, of any proposals of plans, programs or dispositions of general character, or, where appropriate, of its modification or its revision. The information on the proposals must be intelligible and accessible to the public, and provide a right to participate in the decision-making process and to the relevant public authority. Public authorities, pursuant to Law (2006) are also obliged to take account of the public consultations in their decisions. Within the context of IWRM, Law 58/2005 goes on to require the public authorities to publish and make available to the public with regard to each River Basin District information relating to the calendar for the elaboration of plans and the consultations measures to be adopted; a summary of the main issues to be addressed, and the river basin management plan.

#### *6.3.4.3 Access to justice*

A right of access to redress and remedy is fundamental in ensuring that those responsible for water resources management are capable of being held to account. In Portugal, a general right of access to redress and remedy by broad civil society is granted by Art. 20 of the Constitution. Pursuant to this obligation, both the State as an entity and public authorities are subject to joint civil liability for acts and omissions committed during the exercise of their functions (Constitution, 2005).

Free legal aid is available in both jurisdictions. In relation to appeals, in Portugal, these may be made with respect to judicial and administrative decisions, but both are made to the Administrative Courts. The Spanish system is possibly more elaborate, with decisions being challengeable, although the mechanisms for such may differ according to the type of the original judgement. In the Spanish context, restrictions apply with regard to annual income and it appears that these are now set comparatively low – the average salary in Spain is around EURO 23,560. This is almost 50% above the legal aid threshold in Spain (EURO 15,978), which means that only the poorest families will have full access to legal aid.

As regards the timeframe within which judicial decisions must be made, in some instances Spanish decisions must be made within three months. This is not broadly applied however. Information regarding the environment, or IWRM, is generally available in both jurisdictions, except where national security or defence is concerned, or where the information is of an intimate personal nature (Portugal also includes situations where the information is related to an ongoing criminal investigation). In addition, where environmental information has been requested, there are closely defined circumstances where the public authority is not obliged to respond, corresponding to those outlined in the Aarhus convention – for example, where the authority does not hold the requested information, or it is unfinished, and where the request concerns internal communications the release of which is not justified by the corresponding public interest in disclosure. It can also include information that is protected by a legitimate economic interest (where this is protected by Spanish or EU law).

With respect to the implementation of the above, it has been suggested that the separation of the executive and judicial functions in Spain is less than perfect due to the scope for possible Executive influence in judicial hiring where agreement has not been reached by cross-party committees. It also appears that the practical issue of achieving redress through the courts in Spain is hindered as a result of the large deposit demanded of those seeking relief (in the case of criminal cases rather than administrative ones, which are free), and the length of time it can take to reach a final decision, increasing the chances that any relief granted may be redundant. Evidence obtained by the authors during interviews with lawyers and experts in Spain strongly indicated that applicants would have to wait for between three and five years before final judgement would be obtained, after having gone through the full range of appeals. Similar considerations apply in Portugal, with suggestions that the Portuguese have a low level of faith in the justice system as a result and are more inclined to resort to use of the media as a tool for policy change. It is not clear how the question of expenses actually affects the bringing of environmental redress or relief actions, but the figures quoted above, coupled with the protracted legal process would seem to indicate that it is a major consideration – no data has been found regarding the actions that are not taken as a result of lack of finances. Despite the Constitutional Court's recognition of the right to a healthy environment in Spain, it is not in effect a directly justiciable right because it can only be addressed in the context of a breach of the more fundamental rights – e.g. life, information, human dignity.

Aside from questions relating to the length of time involved in achieving a definitive judgement, issues relating to the competence of both the judiciary and lawyers to properly appreciate and adjudicate on questions of water management. Judges in the lower courts can be inconsistent in their verdicts, leading to difficulties with respect to the predictability of outcomes. Interpretation of applicable law was also in some cases unnecessarily strict, resulting in people being more inclined to act beyond the law. The result of this inconsistency is a higher level of uncertainty and unpredictability in judicial decisions, and this could also have implications not only for the number of cases that go to appeal but on the inclination of people to act beyond the law. Again, such experience is found in Portugal, with specific training for lawyers with respect to environmental law being provided by the Environment Institute in an effort to combat what is acknowledged to be a weakness among the judiciary and the legal profession.

### 6.3.5 Challenges in implementing IWRM related law

The above analysis shows that while there may still be gaps in the law related to IWRM and governance in both Spain and Portugal, the transposition of regional laws adopted by the UN ECE and EC has ensured that the fundamental principles are articulated at a national law level.

Potential problems exist with the lack of detailed regulations, the main problem often arises from the lack of physical capacity on the part of the responsible agencies. In the Alentejo region, for example, there is a lack of educated operatives to monitor water use and establish appropriate monitoring systems. Unauthorised water use is acknowledged to take place, but without better monitoring systems, it is difficult to remedy this. Problems have also been encountered with the judiciary balking at applying the full penalties available on transgressors who are brought to court, because the fees are regarded as being too high. The implication is that perhaps the balance between penalties being too low to be credible and too high to be applicable has not yet been found. In the Tagus, enforcement of anti-pollution legislation has been patchy despite the existence of paper mills, industrialised pig farming and tanning upstream from Lisbon. It is also clear that the decision-making processes in Spain and Portugal, and the criteria used for the allocation of water use rights, for example, is not as transparent as it might be. With enforcement and changes to rights being difficult to effect through the courts, problems may begin to mount as users perhaps begin to challenge administrative decisions in ways that they have not done before. Land use is not properly taken into account with respect to flooding, with construction still ongoing within floodplains in Portugal following the recent abortive legislation on the issue.

A wealth of legislation is clearly evident within the Tagus case study. However, the true challenge will be not only ensuring the basic implementation of the law, but also complying with the spirit in which that legislation was adopted.

## 6.4 REFERENCES


Aarhus Convention (1998) Convention on Access to Information, Public Participation in Decision-Making and Access to Justice in Environmental Matters, Jun. 25, 1998 (entered into force Oct. 30, 2001), *International Legal Materials* **38**, 517.

Act (1998) Law 67/98, of 26 Oct., Act on the Protection of personal data, transposing into the Portuguese legal system Directive 96/46/EC of the European Parliament and of the Council of 24 October 1995 on the protection of individuals with regard to the processing of personal data and the free movement of such data, Portuguese Government Official Journal, 122/2000, pp. 247.

Biodiversity Convention (1992) UN Convention on Biological Diversity, 5 Jun. 1992 (entered into force 29 Dec. 1992), *International Legal Materials* **31**, 822.

Climate Change Convention (1992) UN Framework Convention on Climate Change, 9 May 1992 (entered into force 24 Mar. 1994), *International Legal Materials* **31**, 849.

Constitution (2005) Constitution of the Portuguese Republic, 7th Revision, 2005, http://app.parlamento.pt/site_antigo/ingles/cons_leg/Constitution_VII_revisao_definitive.pdf

Decree (2007) Law 226-A/2007, *Portuguese Government Official Journal*, 105/2007, 24–48.

Decree (2006) Law 58/2005, Portuguese Water Law: Transposition of the Directive 2000/60/EC, of the European Parliament and of the Council, *Portuguese Government Official Journal* 249/2005, 7280–7310.

Desertification Convention (1994) Convention to Combat Desertification in those Countries Experiencing Serious.

Drought and/or Desertification, Particularly in Africa, 14 Oct. 1994 (entered into force 26 Dec. 1996), *United Nations Treaty Series* **1954**, 3.

Directive (2003) Directive 2003/35/EEC of the European Parliament and the Council of 26th May, 2003, providing for public participation in respect of the drawing up of certain plan and programmes relating to the environment and amending with regard to public participation and access to justice Council Directives 85/337/EEC and 96/61/EC, *OJ L* 156, 25June 2003, pp. 17–25.

Directive (2003a) Directive 2003/4/EC of the European Parliament and of the Council on Public Access to Information and repealing Council Directive 90/313/EEDC, *OJ L* 41/26.

ECJ (2005) Case C251/03, Commission of the European Communities v. Portugal, Failure of a Member State to fulfil obligations – Failure to satisfy the requirements of Annex I to Directive 80/778/EEC – Art. 7(6)-Water intended for Human Consumption, **29** Sep 2005, *OJ C* 296, 26 Nov. 2005.

ECJ (2006) Case C118/05, Commission of the European Communities v. Portugal, Failure of a Member State to fulfil obligations – Directive 2000/60/EC – Community policy in the field of water – Failure to transpose within the prescribed period, 12 Jan 2006, *OJ C* 60 of 11 Mar 2006.

Gooch, G.D. and Stålnacke, P. (eds.) (2010) The Science-Policy-Stakeholder Interface and Stakeholder Participation. In *Science, Policy and Stakeholders in Water Management*, Earthscan, London.

Law (2006) Law 27/2006, of 18th July, regulating the rights of access to information, public participation in decision-making and access to justice in environmental matters. *Official Gazette* **94**, 1 Feb. 2006.

Law (2007) Law232/2007 on the assessment of the effects of certain plans and programmes on the environment. *Official Gazette* **114**, 15 Jun. 2007, pp. 3866–3871.

Maia, R. (2008) The EU Water Framework Directive implementation in the Iberian context. Paper presented at 13th IWRA World Water Congress, 1–4 September 2008, Montpellier, France.

Ramsar Convention (1971) Convention on Wetlands of International Importance especially as Waterfowl Habitat, 2 Feb. 1971 (entered into force 21 Dec. 1975), 996 UNTS 245.

Rieu-Clarke, A. (2004) Sustainable Use and the EC Water Framework Directive: From Principle to Practice? In *International Law and Sustainable Development. Principles and Practice*. N. Schrijver and F. Weiss (eds.) Koninklijke Brill NV, The Netherlands.

Rieu-Clarke, A. (2010) The Science-Policy-Stakeholder Interface and Stakeholder Participation. In *Science, Policy and Stakeholders in Water Management*. G. Gooch & P. Stålnacke (eds.), Earthscan, London.

Royal Decree (2001) Royal Legislative Decree 1/2001, of 20th July, approving the amended Water Act, *Official Gazette* of 24 July 2001).

Spain-Portugal Agreement (1968) Agreement between Spain and Portugal regulating the use and development of the water resources of the international sections of the rivers Miño, Limia, Tagus, Guadiana, and Chanza and of their tributaries, 29 May 1968 (entered into force 7 Apr. 1969) *United Nations Treaty Series* **1206**, 25.

Spain-Portugal Agreement (1998) Agreement on Cooperation for the Protection and Sustainable Use of the Waters of the Spanish-Portuguese Hydrographic Basin, Nov. 30, 1998, (entered into force Jan. 17, 2000), *United Nations Treaty Series* **2099**, 275.

Stec, S. and Casey-Lefkowitz, S. (2000) The Aarhus Convention: An Implementation Guide, United Nations, Geneva

Watercourses Convention (1997) Convention on the Law of the Non-navigational Uses of International Watercourses, May 21, 1997 (not yet in force). *International Legal Materials* **36**, 700.

WFD (2000) Directive 2000/60/EC of the European Parliament and of the Council establishing a framework for Community action in the field of water policy. *Official Journal L* **327**, 22/12/2000, 0001–0073.

UN ECE ESPOO Convention (1991) Convention on Environmental Impact Assessment in a Transboundary Context, Feb. 25, 1991 (entered into force Sept. 10, 1997), *International Legal Materials* **30**, 800.

UN ECE Helsinki Convention (1992) Convention on the Protection and Use of Transboundary Watercourses and International Lakes, Mar. 17, 1992 (entered into force Oct. 6, 1996), *International Legal Materials* **31**, 1312.

# Chapter 7

# Innovative technology and institutional options in Rainfed and irrigated agriculture in the Tagus basin

*J.M. García-Ruiz, S. Beguería, S.M. Vicente-Serrano, J.I. López-Moreno, N. Lana-Renault and T. Lasanta*

## 7.1 INTRODUCTION

In Spain and Portugal, complex agricultural systems have developed due to contrasting climatic characteristics and difficult topographic conditions. In spite of that, agriculture within these countries has a long history of efficient adaptation to water scarcity and steep gradients. Many innovative agricultural developments, including increases in crop production, the genetic diversity of plantsand the introduction of new crop systems, originate from the Mediterranean (Tonriveris, 2003; Arans, 2004; Thivet and Blinda, 2009). Agriculture in Spain, like in other Mediterranean countries, was traditionally characterised by the contrast between irrigated areas and dry land farming areas. The latter faced many difficulties, the most important being the scarcity of precipitation and their seasonal and inter-annual irregularity.

Most of Spain is defined as dry, since the average annual precipitation is lower than 600 mm within an area of high evapo-transpiration demand. In some areas the average annual precipitation does not exceed 350 or 400 mm, as is the case in the Central Ebro depression, parts of the Duero and Tagus basins and Southeast Spain. For instance the city of Guadalajara, close to the headwaters of the Tagus basin, records on average 418 mm per year, while the city of Toledo, in the middle of the Tagus basin, receives 373 mm per year. Figure 7.1 shows the spatial distribution of the average annual precipitation over the Spanish part of the Tagus basin. Precipitation can also vary considerably from one year to another; humid and dry or very dry cycles can be detected when yearly precipitation is analysed in the long term. Recent studies confirm the occurrence of such cycles and their relation to oscillations in the General Atmospheric Circulation patterns, especially the North Atlantic Oscillation (López-Moreno *et al.*, 2007). Multiannual drought episodes in the Mediterranean area are well known and occur with relatively high frequency (Estrela *et al.*, 2000; Lorenzo *et al.*, in press), resulting in severe problems regarding water availability and management (Garrote *et al.*, 2007).

Seasonality is another severe problem in Mediterranean climates as precipitation tends to occur in autumn and spring, with reduced precipitation during winter and, more notably, in summer. The summer dry period coincides with the highest temperatures. On average, the temperature in July ranges between 18 and 23°C, with daily extremes close to 40°C and this period coincides with the highest water demand for forests and crops. For this reason, dry farming systems were not able to produce food for local populations every year, causing the so-called subsistence crises in past centuries. Consequently, traditional crops under dry farming systems have included cereals (which end their vegetative period at the end of spring or the beginning of summer), vineyards and olive trees (which can support long and intense dry periods

 *Integrating Water Resources Management: Interdisciplinary Methodologies and Strategies in Practice.* Edited by Geoffrey D. Gooch, Alistair Rieu-Clarke and Per Stålnacke. ISBN: 9781843393252. Published by IWA Publishing, London, UK.

due to their powerful root systems). Based on our studies the following chapter focuses on the main characteristics of rainfed agriculture in the Tagus Basin, explains the development of irrigated areas, as well as their innovative technology, and why they offer a promising option for land management in rural areas.

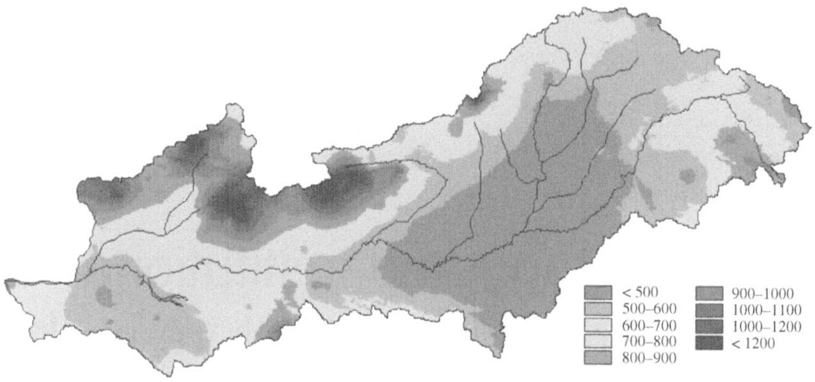


**Figure 7.1** Spatial distribution of average annual precipitation in the Spanish Tagus basin.

## 7.2 RAINFED AGRICULTURE IN THE TAGUS BASIN

Rainfed agriculture in Spain (and in other Mediterranean countries) can be defined as a natural agricultural system in which the only water inputs are supplied by precipitation. Such an approach contrasts with irrigation agriculture, in which a part (sometimes the most important part) of the water inputs are supplied artificially through canals and ditches from rivers and reservoirs. Rainfed agriculture (in Spanish: *secano*) is similar to dry farming agriculture.

In the Tagus basin, rainfed agriculture is the most common form of land use, in spite of low precipitation in the central parts of the basin. The reliance on rainfed methods, however, did not severley affect traditional agricultural methods, as they had been adapted to accomodate the natural conditions of climate and soils. However, crop production was subjected to very high interannual variability and low profitability, due to:

- Scarce average precipitations, particularly in the central part of the basin, where figures of about 400 mm are usual and precipitation shows large variability from one year to another, in such a manner that years with high crop yields can alternate with almost unproductive ones.
- Soils are, in general, relatively thin and poor in organic matter. If plant cover is eliminated, gullies, rills and sheet wash erosion erode the upper soil horizons, particularly during intense rainstorms (De Alba *et al.*, 1998). This is a very important problem, since Mediterranean type crops leave the soil unprotected or scarcely protected during a long period of the year (vineyard from October to April, and cereals from July to March), contributing to a long-term disturbance of soils and compromising their future productivity.
- Relief also shows a large variability. Apart from the mountain, headwater sectors of the tributary valleys, with spatial limitations for agriculture, the rest of the Tagus Basin ranges from horizontal platforms in La Alcarria to undulating, smooth reliefs *(campiñas)* in the southern part of the Madrid region and north of Extremadura. In general, from a topographic point of view, most of the basin indicates favourable conditions for rainfed agriculture.

Rainfed agriculture remained relatively stable for centuries. The characteristics of the local climate only favored the introduction of crops adapted to scarcity of precipitation, high temperatures, intense evapotranspiration and long dry periods during the hot season. This is the reason why cereals, vineyards and olive trees have been the traditional agricultural *trilogy* in Mediterranean environments. Cereals complete their vegetative period just at the beginning of the hot, dry season, while both vineyards and olive trees have their roots and leaves adapted to seasonal drought. These crops have provided subsistence means and even surpluses for local and regional markets since the Roman times. Nevertheless, some changes occurred in rainfed agriculture during the last decades:

- The generalised introduction of machinery and of chemical fertilizers in the Spanish agriculture contributed to a reduction of the area occupied by in-fallow land. Traditionally, cereals cultivation was characterised by the alternation of one year of cultivation and two or three years of repose (fallow), leaving the field to recover nitrogen and other important nutrients. Now, the period of repose tends to be shorter, because the labours with the tractor and plough are more deep, and because of the increasing use of fertilizers. Then for most farmers, the effective area cultivated each year has increased.
- The drastic decrease in the number of inhabitants and farmers in rural areas explains the trend to enlargement of exploitations under rainfed agriculture. This has been the most significant change in the Tagus basin.

The younger and the medium-age farmers bought machinery powerful enough to manage tens of hectares in an extensive way, by buying or renting the land to other landowners. Thus, a natural trend to land concentration increased the average income per farmer, with no intensification of the crop systems nor a changing of the crops.

- Recently, alternative crops were introduced under rainfed agriculture. This is the case of the sunflower, promoted by subsidies from the European Union. Also, in relatively marginal areas, varieties of almond trees resistant to severe drought or to frosts occupy the more and more extent. In the case of vineyards, new varieties have been incorporated to improve the quality of wines and to facilitate the works with machinery. But no other possibilities exist for commercial crops under rainfed agriculture of the Tagus basin, except for some herbaceous crops close to the headwater.

The Agrarian Policy of the European Union also provided new options by subsidizing some crops or activities on traditional rainfed agriculture lands so as to:

- reduce the surpluses in certain crops such as cereals or wine, and increase the production of those crops with deficit or good prices in international markets.
- improve the quality of other products.
- improve the quality and sustainability of the environment.

For these reasons, rainfed agriculture in the Tagus basin is driven by some political and economical measures. Vineyards were affected by a reduction in area due to wine surpluses in the European Union. The objective was the abandonment of the worst lands and the improvement of the quality of wine by introducing new varieties, particularly in Madrid and in the north of Extremadura. Olive trees on the other hand have been favoured by subsidizing not by the production per hectare but by the total extent per farmer. The reasons for this policy were: the existence of an increasing international market for the olive oil; a close relationships between number of olive trees and necessities of manpower, the crop being considered as "social" for this reason; and the idea that olive trees use the water in the soil very efficiently, providing good environmental support during long drought periods.

For years, a very important mainstay of the EU Agrarian Policy was to encourage land abandonment, particularly in fields under dry farming agriculture. The purpose of such a policy was twofold: to reduce cereal surpluses in the European Union by reducing the surface devoted to its cultivation, regardless of the quality of the land; and to improve the quality of soils and the water cycle by ploughing the fields at least once per year. Farmers received a subsidy per hectare abandoned land, and were then obliged to undertake certain tasks in the fields each year. Some marginal lands were abandoned, particularly in areas close to the headwaters of the Tagus Basin (Guadalajara province), with an ageing population.

The EU Agrarian Policy also encourages reforestation in both cultivated and uncultivated areas. Considering that farmland abandonment could result in soil erosion risk, especially during the first years after the retirement of crops, subsidies were also given to those farmers that planted trees (for commercial or conservation purposes) in the fields, in order to improve the water cycle and to reduce soil erosion. This is a promising option in marginal lands, though up to now it has been accepted by a limited number of farmers. In the same way, the cultivation of aromatic plants (thyme, lavender, mint, rosemary, camomile) was also subsidized, but the problem is that farmers are unaware of the commercial outlets for this type of production and furthermore are unfamiliar with the cultivation techniques required, providing a clear example of the need for capacity building activities to achieve a desirable degree of implementation of new policies regarding land use change.

In summary, rainfed agriculture in the Tagus basin faces similar problems to other dry land areas in Spain which include: scarce crop diversity, interannual irregularity of crop yields, and difficulties for introducing new crops. Subsidies have contributed very much to increase the incomes of farmers and the quality of products (especially olive oil and wine). But in spite of these efforts, rainfed agriculture is increasingly becoming a marginal activity in the Tagus basin when compared to irrigation agriculture.

## 7.3 THE DEVELOPMENT OF IRRIGATION AGRICULTURE

Coping with water scarcityin the past included the control of water in small Mediterranean rivers, resulting in the development of irrigation techniques during pre-roman times which revolutionised agricultural practices. Irrigation provided primitive farmers with the capacity to:

- increase the diversity of crops, and then the possibility of a more varied food diet;
- increase crop yields, for example multiplying by several times the production of cereals under dry-farming systems; and
- ensure an intensification of production, thus allowing an increase of population and even a diversification of activities, in such a manner that a part of population could be dedicated to other, non-agricultural labours (artisanal, cultural or political).

The historical development of irrigated lands in Spain is well known. During the Roman period and the Middle-Ages farmers tried to control the small river tributaries, due to the limited capacity of public works to impound the streamflows. Small dams, wells and springs were used to derive water throughout canals and irrigate small areas located on the lower fluvial terraces and flat areas very close to the source water points (Higueras, 1969; García-Ruiz *et al.*, 2003). For this reason, irrigation in Spain mainly affected the small tributaries and not the main rivers. In the case of the Tagus River, some irrigated areas were created very early on in tributaries, following the courses of the rivers and serving local populations or regional markets, as in the case of the Jarama, Manzanares, Guadarrama and Henares rivers. For instance, the Canal of Henares, located in the Henares River, was planned in the 18th century, as are a complex, very old network of irrigation ditches, well known in the whole Tagus basin, using surface water resources (Cabo Alonso, 1960).

The policy on irrigated areas changed in the 20th century, due to technological progress in civil works and an increasing role of the central State Administration, as in other areas in Spain (Lasanta, 2009). New irrigated areas were created in the Tagus River and in some of its main tributaries (Tiétar, Alagón, Jarama, Henares rivers), occupying fluvial terraces but also some plains of structural origin which are elevated with respect to the rivers (Miocene plains in some areas of Extremadura, close to the border between Spain and Portugal). This was possible by constructing up to 75 dams in the Tagus basin, mainly in its tributaries, with different purposes which included: electric power production, agriculture, urban water supply (mainly in the tributaries upstream of Madrid, as in the case of the Lozoya River and the corresponding Canal of Isabel II), or even complex purposes (agriculture and electric power production) (Flores-Montoya *et al.*, 2006). In fact, since the beginning of the 20th century therehas been a clear policy for creating new irrigation areas. The Tagus basin was included from 1933 in the National Plan with the creation of 115.440 hectares of irrigation lands. Since 1956 the Tagus basin has been particularly affected by this policy, due to:

- the increasing demand for electricity in Spain, particularly in Madrid;
- the existence of favourable topographic and lithological conditions in the headwater of the tributaries and also in the main stream; and
- a favourable policy for generating electric power owing to the unification of electric rates for the whole Spain, resulting in the distance between electric generation and consumption places not being a relevant factor.

Some examples of reservoirs in the Tagus basin are the Gabriel y Galán reservoir (924 $hm^3$) in the Alagón River, one of the most important in the Tagus basin; El Vellón (40 $hm^3$) and Valmayor reservoirs in the Guadarrama River headwater; the Tejera reservoir in the Tajuña River; the El Vado reservoir in the Jarama River; and the Burguillo (215 $hm^3$) and San Juan reservoirs in the Alberche River. All these reservoirs are located in the headwaters of Tagus tributaries, affecting the spatial organisation of mountain areas in the Central Range of Spain. But the Tagus River itself was also dammed in several parts, from the headwaters at Entrepeñas (804 $hm^3$) and Buendía (1638 $hm^3$), in the middle course (Estremera, Almoguera and Valdecañas reservoirs, the latter with 1446 $hm^3$ of storage capacity), and at the beginning of the lower course, immediately before the border between Spain and Portugal at the Alcántara reservoir (3,163 $hm^3$, being the largest in Spain and the second largest in Europe), in most cases constructed for the purpose of generating hydro-electric power. Nevertheless, in the case of the Alcántara reservoir, there are some plans to enlarge the irrigated area of the lower alluvial plain of the Tagus River in Portugal (the Ribatejo region). The Arrocampo reservoir (36 $hm^3$) was constructed for the purpose of serving the necessities of cooling the nuclear plant of Almaraz only. Figure 7.2 shows the evolution in the number of reservoirs and also in the water storage capacity of the reservoir in the Tagus basin from the beginning of the 20th century.

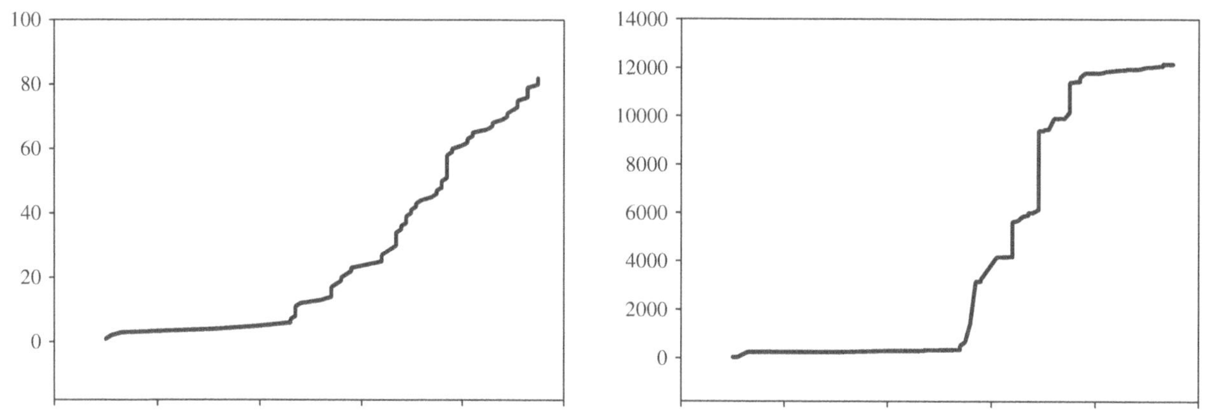


**Figure 7.2** Evolution of the water storage capacity in reservoirs of the Tagus Basin. (Source: Confederación Hidrográfica del Tajo – Tagus Basin Water Authority).

Reservoirs in the headwater of the Tagus River were constructed during the 1960's in order to supply water to the water transfer system installed between the Tagus and Segura basins (García-Yelo, 1997). Prior to their construction, the only way to ensure water supply for irrigation and urban development in the Segura basin in the South-East coast of Spain was by water transfer from the headwaters of the Tagus basin. Due to its location in the driest area of Spain, the Segura basin underwent frequent severe dry periods, with water scarcity threatening the horticultural crops and preventing the development of fruit trees. The extra water supply from the Tagus headwater protected the traditional irrigated areas, allowing the introduction of new commercial crops and the expansion of irrigation to new areas, thus increasing the production of early and expensive crops in Spain and increasing the trading ability of Spain within the agricultural European markets. The Tagus-Segura Canal takes water from the Entrepeñas and Buendía reservoirs and, after 300 km, enters the Alarcón reservoir, in the headwaters of the Segura River.

The water transfer started in 1978, although it never reached the volumes of water forecasted. The original plans estimated a volume of water transfer of about 1100 $hm^3$ per year, but in practice the volume rarely exceeded 500 $hm^3$ per year. This was the reason for new plans to transfer water directly from the Ebro River (located about 500 km northwards) to the Segura River basin, but has finally been rejected by the national parliament. Figure 7.3 shows that there are some fluctuations in the volume of water transfer, with positive years coinciding with humid cycles and the corresponding increase of water storage in the reservoirs, whereas years characterised by water scarcity (i.e., 1983, 1984, 1992–1995) coincide with years of lower water storage (Lorenzo *et al.* in press).

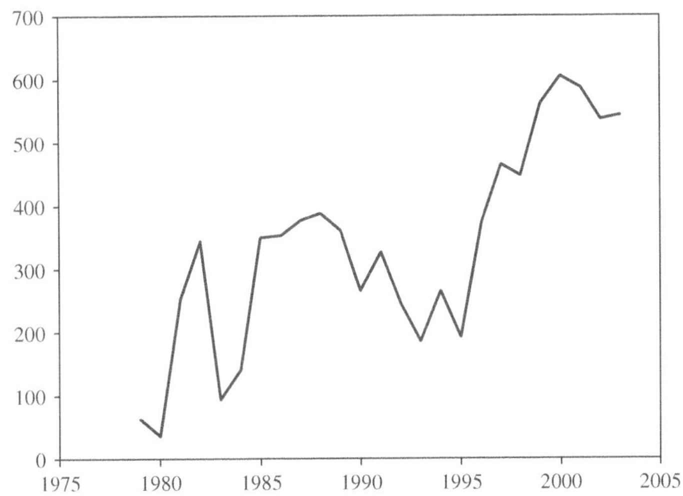


**Figure 7.3** Evolution of water transfers from the Tagus to the Segura Basin (in $hm^3$).

The volume of water transfer must be approved year by year (and even month by month or season by season) by the Spanish Government, according to: the volume of water stored in reservoirs of the Tagus Basin; the amount of precipitation during the wet season; and to the state of reservoirs, soil humidity and previous precipitation in the Segura River basin. Therefore, it is an example of a complex water management system that generates political conflicts between regions. It is interesting to note that the Tagus-Segura water transfer considered in 1971 the necessity to compensate the whole Tagus Basin for the reduction of stream flow and water resources availability in the Tagus River. Compensations included the ending of irrigation areas under transformation and their corresponding infrastructures in the Alagón, Almonte,

Tiétar and Jerte rivers in the Tagus Basin, as well as the encouragement of local irrigated areas and the construction of a water purification plant in the city of Cáceres. It is interesting tonote that the Tagus-Segura water transfer severely affects the water resources of the Tagus basin and the discharge arriving at the border between Spain and Portugal, particularly during periods of water scarcity. For this reason, international agreements are needed in order to ensure a certain level of discharge in the Tagus River, especially in summer, when the water quality decreases significantly. This problem needs to be solved with the management of the Alcántara reservoir, although it is not always achieved (López-Moreno *et al.*, 2009). The Alcántara reservoir was constructed in 1969 to produce electric power, with a large capacity to store water (3162 $hm^3$). Water storage is conditioned by the precipitation characteristics of each year, by the volume of water transferred to the Segura basin and by the volume of water used in irrigated areas upstream of the reservoir. Downstream, the effects of the reservoir can increase the duration and intensity of droughts. Recent studies confirmed that, when water is abundant, the Alcántara reservoir is able to release high discharge during winter and spring while ensuring the hydropower production. In those years, the water release at the foot of the dam enables increasing the downstream summer flow, alleviating the seasonal drought. Conversely on the dry years the managers of the Alcántara reservoir maintain the minimum required reservoir level with the elimination of winter and spring high flows, amplifying in some cases the drought conditions downstream of the dam (López-Moreno *et al.,* 2009). For this reason since February 2000 the management of international waters shared by Spain and Portugal is regulated by a bi-lateral agreement, the "Albufeira Convention" (see Chapter 6). This agreement establishes the minimum flows to be released by Spain to Portugal during normal years, and the necessity of prior communication to Portugal during exceptional floods or during droughts. Although few years have passed to assess the success of the agreement in terms of improving the integrated water resources management, the document is a promising tool to minimize the downstream impacts of the Alcántara dam.

Modern irrigation explains the general trend to extend agricultural systems in irrigated areas, although it is not the only factor. There are huge differences between the old traditional irrigation areas and the modern ones, which include (Higueras, 1969):

- Traditional irrigated areas occupy small areas, the fields and properties are smaller, and the soils are deep, rich in organic matter and very productive. They mainly serve for self-consumption and the local markets.
- Modern irrigations occupy large areas and the fields are as large as those under dry farming. Soils are poorer in organic matter and they only allow one crop per year, and cultivation is linked to national or even international markets (sugar beet, cotton, at the beginning, and fodder, tobacco, tomato, later, according to market preferences and prize fluctuations).

Many other plans existed for the construction of new reservoirs in the last three decades of the 20th century, some of them very large, but the hydraulic and environmental policy changed and most of them were never built. For instance, in the Tiétar River a new reservoir was constructed in the nineties, the Rosarito dam, with a storage capacity of 100 $hm^3$, capable of irrigating about 15.000 hectares. Two new reservoirs were also planned in the same Tiétar River, the Arenas de San Pedro and Monteagudo reservoirs, with a total capacity of 880 $hm^3$, allowing the enlargement of the irrigated area, and the control of relatively frequent floods. Nevertheless, political and ecological arguments have led to the rejection of the construction of the first of the dams and, presumably, would discard or reduce the storage capacity of the second one.

In summary, the Tagus River basin was affected throughout the 20th century by the construction of a large number of reservoirs. Some of the main irrigated areas locate in fluvial terraces of the Tagus River, downstream of Toledo, following a relatively narrow lane spatially restricted by the morphological characteristics of the valley. Nevertheless, the most important irrigated areas occupy relatively large sectors of the tributaries, where there is favourable terrain and abundant water resources. Thus the Tajuña, Henares, Jarama and Guadarrama rivers organised new irrigated areas in fluvial deposits and enlarged the relatively small areas that were irrigated centuries ago. But by far the most important transformation occurred in the westernmost tributaries of the Spanish sector of the basin (Bueno-Hernández, 1996), where a total of 63.404 hectares were transformed into irrigation during the 1960's and 1970's. This transformation affected three areas:

(1) the Tiétar system, with 20.235 hectares;
(2) the Alagón system, with 33.799 hectares; and
(3) the Arrago system, with 9.370 hectares.

Figure 7.4 shows the spatial distribution of irrigated areas in the Spanish sector of the Tagus Basin.

**Figure 7.4** Distribution of irrigated lands in the Spanish sector of the Tagus River basin. (Source: Confederación Hidrográfica del Tajo – Tagus Basin Water Authority).

The total irrigated area in the Tagus basin is 231.000 hectares, divided into 115.000 hectares for public initiative and 116.00 hectares for private initiative. It is interesting to note that there are several areas without irrigation and others that are so small that cannot be represented in the map. These areas are located in the headwaters, due to terrain and climatic reasons, and the left side of the western sector, due to the scarcity of water resources, the low quality of the soils (extremely thin) and the uneven relief, with ravines incising into the pediment. In general the left side of the basin shows a lower development of irrigation, and this is mainly due to the poor discharges of the tributaries from that side. In any case, the percentage of surface covered by irrigated lands (3.5%) is very low as compared to other Spanish basins (Table 7.1).

Table 7.1 Percentage of surface that represent some land cover classes in the Tagus basin. (Source: Confederación Hidrográfica del Tajo – Tagus Basin Water Authority).

| Land-cover | Percentage |
|---|---|
| Urban and industrial areas | 1.67 |
| Non-irrigated arable lands | 25.84 |
| Irrigated lands | 3.47 |
| Vineyards and olive grooves | 3.79 |
| Pastures | 11.73 |
| Mixed agricultural/natural vegetation | 12.89 |
| Forests | 15.82 |
| Shrubs and maquia | 24.80 |

## 7.4 PROMISING OPTIONS IN IRRIGATED AGRICULTURE

Irrigation in the Tagus basin is a promising option for farmers compared to dry farming agriculture in Central Spain. In general, irrigation increases by several times the crop yields and gives opportunities of introducing new crops which are more water-demanding. For this reason, irrigation agriculture is considered as a very positive factor for retaining population in rural areas and also for increasing the incomes of farmers, encouraging the location of other economic activities. The new irrigated areas show increasing economic activity and give farmers the opportunity of improving their living conditions. This section shows the most interesting promising options for the development of irrigation agriculture in the Tagus basin.

According to estimations of the Hydrological Plan of the Tagus Basin, the total amount of water resources in the Spanish sector of the Tagus basin are 12.230 $hm^3$ on average per year. Of that figure, 1.539 $hm^3$ correspond to groundwater resources, particularly important in the Tajuña basin (545 $hm^3$), the Madrid-Talavera aquifer (401 $hm^3$), the Tiétar basin (169 $hm^3$) and La Alcarria (122 $hm^3$). In the case of surface water resources, the average discharge of the Tagus River in Alcántara is 188 $m^3 s^{-1}$, and that of the main tributaries are: the Tiétar River, 59 $m^3 s^{-1}$ in Torrejón, the Alagón River, 29,8 $m^3 s^{-1}$ in Coria, and the Jerte River, 14,9 $m^3 s^{-1}$ in Galisteo.

This means that there are abundant resources in the whole basin and that, under an adequate policy of water storage and management, these resources are enough to satisfy the basin's water demand. Nevertheless, it is important to remember a number of issues: that the main purpose of most reservoirs, but mainly the largest ones, is electric power generation and irrigation is a secondary use; that the complex network of reservoirs located in the northern part of Madrid exclusively serves and supplies water to the domestic and industrial sectors (Cubillo and Ibáñez, 2007); and part of the water resources are diverted to the Segura basin, to supply water to the irrigated areas in southeast Spain. Besides, the transboundary agreements between Spain and Portugal obligates these countries to maintain a certain discharge level in the Tagus river, particularly during dry periods, when water demand is at its highest level for irrigation purposes.

Irrigation agriculture represented deep changes in some areas of the Tagus basin. For centuries, irrigation was a strategy to ensure subsistence and, in the best cases, to increase incomes from the local markets. But the new irrigated areas face other challenges, including:

- The incorporation of traditional rural areas to national and international markets. This represents the cultivation of relevant products as an alternative to the traditional Mediterranean crops (cereals, olive trees, vineyards). The new crops are mainly corn, sunflower, tomato, onion and melon, together with tobacco. The productivity is relatively high in the irrigated areas close to Talavera de la Reina and also in the Alagón and Tiétar irrigations. For instance, corn production is about 13,000 kg per hectare, peas, 5000 kg per hectare, lucerne, 20,000 kg per hectare, lettuce, 50,000 kg per hectare, melon, 35,000 kg per hectare, pepper, 40,000 kg per hectare. Of course, the productivity of traditional cereals also increases: wheat, 5500 kg per hectare, and barley, 5800 kg per hectare instead of 1500 kg per hectare under dryland farming systems (Ortega *et al.*, 1999).

- The necessity of maintaining or even increasing the population in the villages as a main purpose for rural development. It is well known that rural population decreased intensively in Spain during the 1960s and 1970s, and that farmers represent a decreasing number in relation to the total active population in Spain. In the case of the Tagus basin irrigated areas, the proportion of farmers is about 12% in the Toledo province, close to 20% in Extremadura, and less than 2% of active population in the region of Madrid. This purpose has been achieved only in a partial way. Unfortunately, irrigation of large areas arrived too late to counteract the natural demographic evolution of the inner Spain. When irrigation arrived to the Alagón and Tiétar valleys (Extremadura) an important proportion of young farmers had already migrated to the cities. Then, the new irrigation techniques faced a very important problem: the scarce number of young farmers. Nevertheless, villages with new irrigated areas show a better demographic composition, with a higher number of young people and women.
- The location of new companies devoted to the transformation of food products (agroindustry) and other activities. The existence of irrigated areas increases the possibilities of a greater economic complexity, including: machine and repair shops, factories of transformation of corn and lucerne into fodder, canning industries (particularly important in the case of tomato), commercialisation companies, transport companies, and, around them, other tertiary activities (insurances, diverse shops due to the increase of incomes, and so on) and even building and restoration of houses. This represents at least a stabilization of the population and a slight increase in the best cases, particularly in the traditional central towns.

In fact, central towns stabilized and increased their population whereas the small villages stabilized or even reduced the number of inhabitants because of the progressive reduction in the number of farmers and the progressive concentration of the rural properties. This is a very interesting topic, in that, modern irrigated areas affected both the small and large landowners, but the natural evolution tended to reduce the number of small landowners due to their inability to compete in a relatively open market. In fact, the evolution of the land property in Extremadura shows a clear reduction in the smallest properties and an increase in the number of the largest ones.

An interesting case in the evolution of small and large properties is the irrigation of the Tiétar valley. Small properties tended to introduce a complex mosaic of crops, trying to grant the subsistence and to obtain incomes while minimizing the risk. Thus, in a farm of about 10 hectares or less, the farmer cultivates a permanent garden throughout the year, as well as lucerne, cotton, corn, peanuts, wheat and barley, sometimes accompanied by a small number of pigs. Large properties show a more homogeneous crop system, with complex grasslands (up to 50% of the total), tobacco, fodder corn and oats, accompanied by relatively extensive livestock; that is, in the larger properties (frequently higher than 100 hectares), a complementation between cultivation and livestock has been incorporated, and the livestock products are the main source of economic incomes (Rodríguez-Cancho, 1983).

The incorporation of Spain into the European Union on January, 1, 1986 represented new promising options and opportunities for the irrigated areas in the Tagus basin, as in the rest of Spain (Pérez-Díaz, 2006). It must be taken into account that in 1986 irrigation agriculture in the Tagus basin was affected by an unbalanced distribution of the land property, a scarcity in the presence of agrarian industries close to the places of production, an absence of commercialisation channels, a lack of farmer associations, and the predominance of cereals, olives, fruits and a few number of vegetables among the exported products. The Agrarian Community Policy tried to solve these problems, and, in fact, a deep transformation in the irrigation agriculture of the Tagus basin can be observed more than 20 years later. The most interesting political decisions include the following:

- Quotas for milk production, affecting particularly (and negatively) the small farmers.
- Establishment of aids for livestock, mainly sheep, goats and cows for meat production.
- Modernization and re-dimensioning of the land property, as a consequence of a reduction in the number of farmers.
- Aid for the rejuvenation of farmers (certainly with little success), encouraging the substitution of the parents or even grandfathers by the young members of the family.
- Advances in technical training, in order to fight against the scarce qualification of the farmers. With this purpose training courses have been funded by the public administrations explaining new crop systems, irrigation procedures, and the introduction of new crops.
- Aid to industrialisation and commercialisation.
- Promotion of associations, that runs against the distrust of the farmers due to their aged structure.

At present, subsidies from the European Union result in a clear increment of the agrarian incomes, in such a manner that 44% of such incomes are supplied by the European Union. Nevertheless, it is important to take into account that 80% of the aid goes to 20% of the landowners because of the unbalanced structure of the property.

One of the most promising options for the present and near future is a reduction in water consumption per hectare (Ortega *et al.*, 2005). The existence of multiple water users (in particular the Tagus-Segura water transfer) and the enlargement of the irrigated areas in the Tagus basin itself have reduced the availability of water during scarcity periods.

Furthermore, the improvement of commercial networks encourages the introduction of highly water-consuming crops, in substitution of cereals. Recent studies estimated water consumption for different crops in the Tagus basin. For instance, in the Hydrological Plan of the Tagus Basin, cereals would receive an endowment of 3000 $m^3$ per hectare, sunflower, 4500, corn, 5500, potato, between 4600 and 4900, lucerne, between 7000 and 8000, melon, between 3700 and 4200, tomato between 3600 and 4100 and onion between 4500 and 5800 $m^3$ per hectare. That means that in some cases an important effort needs to be made in order to reduce water consumption, above all in corn and lucerne, but also in onion and sunflower crops.

Several studies confirmed that irrigated areas in the Tagus basin (as in other areas of Spain) are over-irrigated because of the low price of the water (Berbel and Gómez-Limón, 2000), resulting in (a) an excess of water consumption, (b) problems of $NO_3$ pollution of the groundwater, and (c) existence of effluents from the irrigation areas which contribute to the pollution of surface waters. Thus, an effort to improve the water efficiency needs to be made in order to reduce water consumption and pollution (Ortega *et al.*, 1999). Usual irrigation methods are based on gravity systems (sometimes surface furrow irrigation), with high losses in the distribution ditches and pipes. During the last few years new irrigation technologies (sprinkler systems and drop irrigation) became more common, reducing water consumption. New methods are now locally tested south of Madrid in corn fields in order to improve the irrigation schedule, based on increasing water storage to field capacity in the 0–50 cm soil layer, resulting in 15% water saving without reducing the crop yields (Román *et al.*, 1999).

Finally, one promising option in the development of irrigated areas in the Tagus basin is the transformation of small dry farming areas into irrigated ones. This is a strategy different to the one prevailing until 20 years ago. During the second half of the 20th century, the irrigation strategy was based upon the creation of very large irrigated areas, including even the construction of new villages with the displacement of populations from other regions. Now the interest is to create new irrigated lands of small size, linked to villages and favouring:

- the stabilisation of the local population, throughout the increase in the number of local employments;
- the increase in productivity;
- the improvement of work conditions (increase in the size of plots by concentration, construction of new pathways improving accessibility to the fields); and
- water saving by using new irrigation systems and training of local populations.

This strategy is being used in different sectors of the basin, particularly close to the headwater area. One of the most interesting examples is that of the village of Bornova in the Guadalajara province, where 2500 hectares of new irrigated land were created since 1993 in the Henares Valley and its tributary, the Bornova Valley (Estévez-Adán, 1996). Nevertheless, two problems must be mentioned. Productivity has certainly increased, although in a limited proportion. For instance, barley under dry farming system yielded about 1800 kg per hectare, whereas under irrigation yields 3000 kg per hectare. The difference is important for the modest traditional incomes in the area, but is comparatively much lower than in other irrigated areas of the basin. In addition, the traditional crops (wheat, barley) are also prevailing in the irrigated fields because of climate and soil conditions and the aged demographic structure of the local population. In summary, the different examples and situations of irrigated areas in the Tagus basin confirm the large variability of environmental and social conditions from the headwaters to the border between Spain and Portugal.

In the headwaters (La Alcarria in the Guadalajara province), the climate neither favours crop diversification nor a strong increment in crop production, and the population is so extremely aged as to limit the possibilities of introducing new systems and crops. The strategy has been the creation of small irrigated areas aiming at fixing the local population and at increasing the household incomes. Westwards the climatic conditions are more favourable to irrigation, even allowing two crops per year (for instance, garlic and lettuce, lettuce and beans, lettuce and cauliflower, or lettuce and late potato), more diverse and productive crops (in part, green vegetables) and the existence of a commercialisation network due to the neighbouring presence of Madrid and Toledo. This is the case of the central Tagus Valley, close to Talavera de la Reina or the tributary valleys south of the Madrid region (Jarama and Henares valleys).

In Extremadura, close to the border between Spain and Portugal, larger irrigated areas were created in the Alagón and Tiétar valleys. There, the climate is favourable for diversification of crops and there is still a certain level of population, although the main markets are away and the commercial networks are still incipient. For this reason more extensive irrigated crops prevail including corn, lucerne, and grasslands as well as other somewhat more intensive (tomato, pepper), directed to local commercialisation in agroindustries.

This demonstrates the possibilities and promising options of irrigation in the Tagus basin, with diverse environmental, social and market conditions. For this reason it is very important that both the State and regional authorities be sensitive enough to that diversity in order to improve adequately the basic conditions for agriculture development.

## 7.5 CONCLUSIONS

In spite of the large promising options for irrigated lands in the Tagus basin, some problems remain to be solved, which can be a direct or indirect source of conflicts. One of the main issues is related to the planning of water concessions to satisfy the water demand for different uses, since irrigation must compete with other industrial and urban water uses. Usually concessions for irrigation are overestimated, partially due to inefficiency in the irrigation systems.

There are major irrigation challenges in the Tagus basin. Such challenges relate to the quantity of water, mainly as a consequence of the frequent water scarcity during the dry years, but also with the water quality, since there is a specific problem related to water quality once the river crosses Madrid, and in some areas (Vega del Tajo) agriculture strongly depends on water quality ("scientific agriculture").

A large effort is being developed to improve the efficiency of irrigated lands in the Tagus basin, and one of the priorities of the Tagus Management Plan is to increase the irrigation efficiency. Thus, priority will be granted to the concessions that save water and encourage the sustainable use of groundwater resources (Martín de Santa Olalla *et al.*, 1999). The improvement and consolidation of the present irrigated land is also a priority with the purpose of increasing their competitiveness, favouring the re-use of water and developing a strict control of the water use.

Improving the efficiency of irrigated lands should also include the use of crop varieties better adapted to the climatic conditions and the development of education programs leading to the establishment of modern irrigation techniques and the improvement of applied research regarding soil-plant-climate relationships and crop production within the basin.

Although several uncertainties remain regarding the future of agriculture, it is clear that the benefits reaped from irrigated lands go beyond a simple economic sector. Irrigated lands are not only highly necessary for the maintenance of ecological/integrated agriculture, based on the implementation of agro-environmental measures that include participatory and integrated management but also support local communities, in order to avoid rural abandonment.

## 7.6 REFERENCES


Arans, J.I. (2004) The problems of sustainable water use in the Mediterranean and research requirements for agriculture. *Annals of Applied Biology* **144**, 259–272.

Berbel, J. and Gómez-Limón, J.A. (2000) The impact of water-pricing policy in Spain: an analysis of three irrigated areas. *Agricultural Water Management* **43**, 219–238.

Bueno-Hernández, F. (1996) El río Alagón: Pasado, presente y futuro de sus obras hidráulicas. *Revista de Obras Públicas* **3356**(49).

Cubillo, F. and Ibáñez, J.C. (2007) Drought management in the urban water supply system of Canal de Isabel II. *Options Méditerranéennes, Série B: Études et Recherches* **58**, 407–415.

Cabo Alonso, A. (1960) Colonización y nuevos regadíos en las Vegas del Tajo. *Estudios Geográficos* **31**, 354–359.

De Alba, S., Benito, G.Y. and Pérez González, A. (1998) Erosión de suelo en episodios de lluvia de elevada intensidad versus episodios de moderada y baja intensidad y elevada frecuencia en ambientes semiáridos. En *Investigaciones recientes de la Geomorfología española*. Gómez Ortiz, A., Salvador, F., Schulte, Y.L. and García Navarro, A. (eds.), Granada, Sociedad Española de Geomorfología, pp. 483–491.

Estévez-Adán, V. (1996) Transformaciones agrarias e impacto ambiental: nuevos regadíos y nuevos paisajes en la Vega del Bornova (Guadalajara). *VIII Coloquio de Geografía Rural*, Zaragoza, AGE-Universidad de Zaragoza, pp. 439–453.

Estrela, M.J., Peñarrocha, D. and Millán, M. (2000) Multi-annual drought episodes in the Mediterranean (Valencia region) from 1950–1996. A spatio-temporal analysis. *International Journal of Climatology* **20**(13), 1599–1618.

Flores-Montoya, F.J., Liébana del Pozo, G., De Ortiz-Andrés, M.A. and Mora-Colmenar, J. (2006) Demographic impact of the regulation of surface waters. The case of the public irrigated areas in the Tagus River basin. *Proceedings of the International Symposium on dams in the societies of the 21st Century*, Vol. 1, pp. 353–355, Barcelona.

García-Ruiz, J.M., Lasanta, T. and Biarge, F. (2003) *Agua, tierra y paisaje: Complejidad y diversidad en el territorio de Riegos del Alto Aragón*. Confederación Hidrográfica del Ebro y Comunidad General de Riegos del Alto Aragón, 164 pp., Zaragoza.

García-Yelo, J.J. (1997) Situación actual del tasvase Tajo-Segura. *Revista de Obras Públicas* **3365**(7).

Garrote, L., Iglesias, A., Moneo, M., Garrido, A., Gómez, A., Lapeña, A., Benbeniste, S., Cubillo, F. and Ibáñez, J.C. (2007) Application of the drought management guidelines in Spain. *Options Méditerranéennes, Série B: Études et Recherches* **58**, 373–406.

Higueras, A. (1969) La agricultura de regadío en España. En: *Miscelánea a Canellas*, Zaragoza, Universidad de Zaragoza, pp. 585–630.

Lasanta, T. (2009) Cambios de función en los regadíos de la cuenca del Ebro: un análisis del papel de los regadíos a lo largo del tiempo. *Boletín de la A.G.E.* **50**, 81–110.

López-Moreno, J.I., Beguería, S., Vicente-Serrano, S.M. and García-Ruiz, J.M. (2007) Influence of the North Atlantic Oscillation on water resources in central Iberia: Precipitation, streamflow anomalies, and reservoir management strategies. *Water Resources Research* **43**, W09411, doi:10.1029/2007WR005864.

López-Moreno, J.I., Vicente-Serrano, S.M., Beguería, S., García-Ruiz, J.M., Portela, M.M. and Almeida, A.B. (2009) Dam effects on droughts magnitude and duration in a transboundary basin: The Lower River Tagus, Spain and Portugal. *Water Resources Research* **45**, W02405, doi:1029/2008WR007198.

Lorenzo, J., Vicente-Serrano, S.M., López-Moreno, J.I., Beguería, S., García-Ruiz, J.M. and Cuadrat, J.M. in press. The impact of climatic droughts and water management on different hydrological systems in the headwaters of the Tagus River (Central Spain). *Water Resources Management*.

Martín de Santa Olalla, F., Brasa-Ramos, A., Fabeiro-Cortés, C., Fernández-González, D. and López-Córcoles, H. (1999) Improvement of irrigation management towards the sustainable use of groundwater in Castilla-La Mancha, Spain. *Agricultural Water Management* **40**, 195–105.

Ortega, J.F., De Juan, J.A., Tarjuelo, J.M. and Merino, R. (1999) Modelo de ayuda a la toma de decisiones sobre el manejo del agua de riego. Aplicación al análisis de la agricultura de regadío en la Vega del Tajo (Toledo). *ITEA* **95**, 149–178.

Ortega, J.F., De Juan, J.A. and Tarjuelo, J.M. (2005) Improving water management: The irrigation advisoty service of Castilla-La Mancha (Spain). *Agricultural Water Management* **77**, 37–58.

Pérez-Díaz, A. (2006) La PAC y los reajustes del campo extremeño. XIII Coloquio de Geografía Rural, Jaén, AGE-Universidad de Jaén, pp. 200–213.

Rodríguez-Cancho, M. (1983) Paisajes nuevos en la provincia de Cáceres. *VIII Coloquio de Geógrafos Españoles*, Barcelona, Universidad de Barcelona, pp. 534–540.

Román, R., Caballero, R. and Bustos, A. (1999) Field water drainage under traditional and improved irrigation schedules for corn in Central Spain. *Soil Science Society of America Journal* **63**, 1811–1817.

Tonriveris, H. (2003) Agricultural land use change and sustainable use of land resources in the Mediterranean region of Turkey. *Journal of Arid Environments* **54**, 553–564.

Thivet, G. and Blinda, M. (2009) Una revisión de las estrategias hidráulicas, en *Toramed, Nuevas perspectivas para el desarrollo rural en el Mediterráneo*. Centro Internacional de Altos Estudios Agronómicos Mediterráneos, Ministerio de Medio Ambiente y Medio Rural y Marino, 75-III, Zaragoza.

# Chapter 8

# IWRM status in the Tungabhadra sub-basin

*S. Manasi, I. Nesheim, K.J. Joy, S. Paranjape, U.S. Nagothu, K.V. Raju and N. Latha*

## 8.1 INTRODUCTION

The Tungabhadra River is both a tributary of a larger river, the Krishna, and a transboundary or inter-state river shared by three Indian States of Karnataka, Andhra Pradesh and Maharashtra. The concept of IWRM in Tungabhadra is relatively new and water management so far has been carried out mainly at state level, thus Karnataka and Andhra Pradesh each have their own set of different water governance regimes – water policies, laws, institutions and practices in place, with relatively little interaction between or amongst them. Water is viewed both as a social and an economic good in the basin, through various policies and measures. Water rights is a strongly emerging issue, as water rights are tied to land rights, creating access problems for those who are landless. This has substantial consequences for the vast majority of people as agriculture is the main source of employment, with irrigated agriculture being of primary importance to the economy. IWRM is mentioned in a few policy documents like the National Water Policy and the Karnataka State Water Policy and there are a few small-scale initiatives that could serve as a starting point for an integrated approach. A recent interesting move in Andhra Pradesh water resources management is the establishment of multidisciplinary teams at the level of developing projects; this can be regarded as a small step in the direction of IWRM, though under the umbrella of 'change management'. Similarly, there are initiatives to protect the river ecology at different points, through local institutions and individual innovative farmers. Additionally, the practice and promotion of water saving technologies is evident in several areas. This chapter evaluates to what degree IWRM principles have been followed in light of evolving strategies and methodologies for improved IWRM. The main message is that an overall integrated plan is needed to envisage how transformation can be achieved within a basin wide management approach. Some of the drawbacks in the basin are presented in Table 8.1.

©2010 IWA Publishing. *Integrating Water Resources Management: Interdisciplinary Methodologies and Strategies in Practice.* Edited by Geoffrey D. Gooch, Alistair Rieu-Clarke and Per Stålnacke. ISBN: 9781843393252. Published by IWA Publishing, London, UK.

**Table 8.1** Some important IWRM issues in the Tungabhadra basin.

| Institutional issues | Stakeholder issues | Environmental issues |
|---|---|---|
| Lack of inter-sectoral and inter-departmental integration or coordination | Absence of a river basin organization | Non-availability and access to reliable data at appropriate scales to the public, which is a pre-requisite for informed participation |
| Absence of a nested institutional framework and a river basin organization | Lack of institutional space for stakeholder participation in governance issues except for efforts towards participatory irrigation management | No integrated management plans as per IWRM principles for the basin though the national and the state level water policies specify them |
| Lack/Poor coordination across stakeholders and institutions in spite of competing water demands and conflicts | Adequate attention is not given for training and capacity building of stakeholders. | Lack of information on pollution and its impacts on ecology, environmental flows and health |
| Lack of basin wise information and data base | Marginal stakeholders are neglected affecting livelihoods | Lack of holistic approach in addressing pollution issues |
| Lack of proper allocation norms and procedures for different sectors and different users | Poor interaction across stakeholders | Lack of integration across different scales (like no integration of ground water and surface water, no integration of tanks and other larger surface water bodies like dams and reservoirs) |

## 8.2 THE CONTEXT OF THE TUNGABHADRA BASIN

The Tungabhadra originates in the Western Ghats and has a catchment area of 71,417 km$^2$, of which 57,671 km$^2$ (nearly 80% of the basin) lies in Karnataka (Figure 8.1). In general terms, the biophysical context of the Tungabhadra basin can be divided into two parts. The upper catchment, located mainly in the state of Karnataka, receives much higher rainfall than the middle, lower portion of the catchment, located mainly within the state of Andhra Pradesh. The average annual rainfall in the catchment is 1200 mm. According to official statistics, farming dominates land use/land cover within Karnataka state as of 2004–2005, accounting for more than 55% of the surface area. Other cultivated areas such as trees and groves, fallow land and fallow cover up to 12.5% of the territory. Forests and natural vegetation cover 16% of the area, and around 5% is used as permanent pastures. 11% of the territory is not available for cultivation or for natural vegetation. The area is dotted with decentralized, localized water harvesting systems known as 'tanks' (see Chapter 9). There are also mining activities within the area. The middle and lower portion of the basin, is characterized by lower rainfall, drought conditions and a sharp delineation between rain-fed and canal irrigated areas, served by the large dams on the Tungabhadra. Here cropland accounts for 38% of the territory, with a higher proportion of fallow land (16%) and non-cultivable land (17%) due to the harsher climatic conditions in the area. The forest or natural vegetation areas cover 23% of the territory.

Agriculture is the major occupation in the region and 80% of the population depends on agriculture for their livelihoods. Irrigation is provided through canal systems; however, in the rain fed regions farmers also extract ground water through bore wells. The major crops grown are paddy, jowar, sugarcane, cotton and ragi (finger millet). Although the area is better suited for semi-arid crops, such as millets, pulses etc., the major crops grown are the high water demanding crops such as paddy and sugarcane. The area occupied by these high water-demand crops is also increasing. The spread of water-intensive cultivation methods throughout the basin has dramatically altered the water balance, leading to major conflicts between water use for cash crop cultivation and staple food production on the one hand, and irrigation, industrial and drinking water needs on the other. There is a significant distinction between those farmers with access to land and irrigated water, and those without access, as those with access to irrigation can obtain higher incomes. The inequity of this situation is reflected in the conflicts between upstream and downstream users.

A change which has made the situation of competition for water even more complex during the last two decades is the increasing number of expanding small towns and industrial areas. While industrialization and growing urban areas mean improved standards of living for some, the same activities can also increase the conflict levels between different water users in the basin. It seems that there has been an inability to keep pace with the ensuing development and provide sanitation and water supply infrastructure to resource poor sections of the communities, both in small towns and in rural areas. Population pressure and increased urbanization have added to the abovementioned problems. With increasing demands from industry and urban areas for water, conflicts are arising between users, for instance, agriculture vs. industry. The socio-economic aspects of the case are of very high relevance where the differences in standards of living between the various social classes are tremendous and the country is in the middle of an accelerating change.

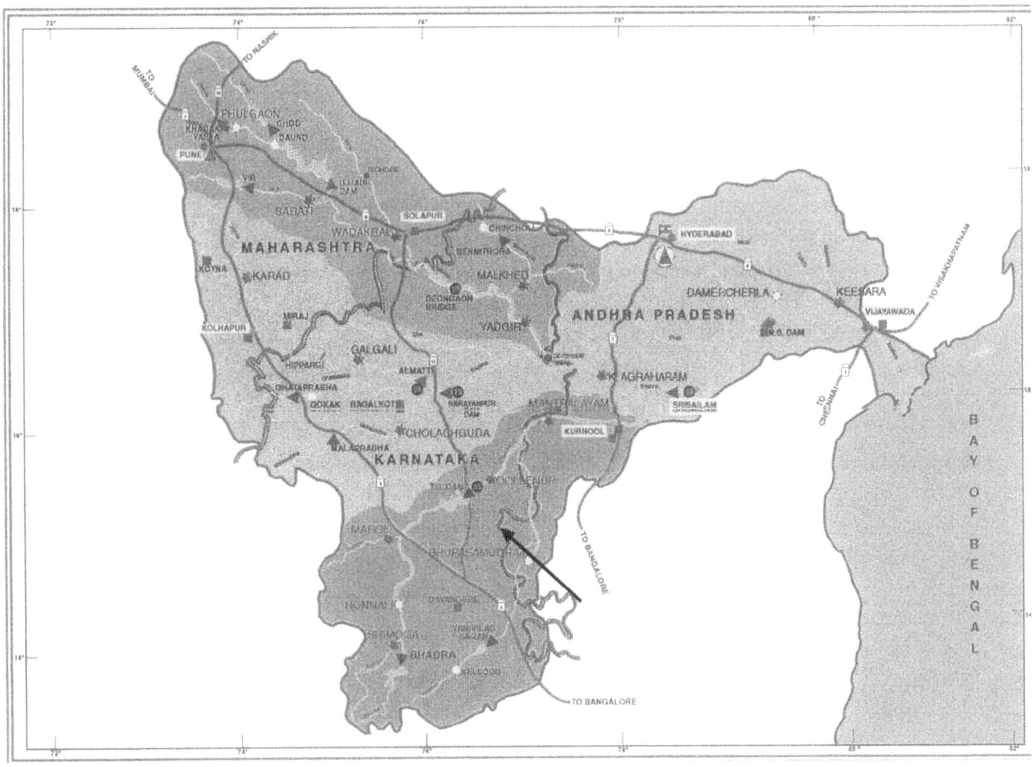


**Figure 8.1** Location of the Tungabhadra sub-basin, India (http://waterresources.kar.nic.in/river_systems.htm#map).

The Tungabhadra sub-basin (TBSB) is a politically sensitive basin due to re-negotiation processes between the riparian states concerning water allocation. Conflicts within and across sectors are common, apart from interstate disputes, due to the transboundary (between Indian states) nature of the river that flows across two states. In the TBSB, there has been no regulatory authority for water, plus no clear or enforceable water entitlements and water rights for the Tungabhadra Basin as a whole. Individual rights to both surface water and groundwater are recognized only indirectly through land rights.

## 8.3 AN ASSESSMENT OF IWRM WITHIN THE TUNGABHADRA

The focus of the following sections will be on five selected components of IWRM, which are vital to the overall goal of sustainable development. These five components are: (i) protection of the catchment and the environment; (ii) measures to ensure efficient and equitable use of water; (iii) institutional analysis and stakeholder involvement in water management; (iv) capacity-building; and (v) transboundary issues (see Chapter 1).

### 8.3.1 Protection of the catchment and the environment

The protection of the catchment and the environment requires integrated management, including the conservation and protection of aquatic and water related biodiversity resources. Yet a basin management plan is still missing for the Tungabhadra, despite national (National Water Policy, 2002) and state policy (Karnataka State Water Policy, 2002) having included IWRM as one of the objectives. The protection measures in TBSB are implemented mostly through the measures implemented by the State Forest Department, through protected areas, deforestation activities and enforcement. Pressures on land for agriculture linked to competing water uses across sectors have influenced land use significantly and in turn have impacts on the environment and people's livelihoods. Runoff from agricultural lands results in water logging, desertification, salinization, and erosion that affect the irrigated areas and water quality. Tillage or ploughing activities results in sedimentation, which carry phosphorus and pesticides, increasing siltation and loss of habitat. Using extensive fertilizers affects the agricultural fields resulting in runoff of nutrients, especially phosphorus, leading to eutrophication causing taste and odor in public water supply, excess algal growth leading to deoxygenation of water and fish kills. Across the Tungabhadra River Basin, fertilizer consumption has increased to 700 tonnes in 2001–05 from 510 tonnes in 1995–96 indicating nutrient loss in the soil. The use of fertilizers has also affected ground water quality in the basin (Fazi *et al.,* 2009).

Pollution is becoming a problem from both point and non point sources. The principal sources of pollution in the Tungabhadra river basin are: (a) run-off from agricultural fields; (b) industrial effluents; (c) sewage from urban settlements; (d) mining activities; and (e) over exploitation of ground water. Domestic and industrial pollution, combined with deforestation, use of pesticides and fertilizers have affected water quality extensively making water unfit for drinking. Industries are another key source of river pollution. The Tungabhadra River is the major source of water for about 77 large-scale industries located on its bank, out of which, 27 are functioning and 50 are under implementation. The major types of industries are iron and steel, paper and pulp, chemical and sugar. Apart from large-scale industries, there are 2543 small-scale industries as of 2006–07. The mining of iron ore at Kudremukh and manganese ore in Sandur in the upper catchments of Tunga and Bhadra has seriously affected the stability of the catchmentleading to severe soil erosion and silting of several small reservoirs, traditional tanks and the Tungabhadra reservoir, thus conflicting with irrigation needs. These mining activities decrease the water depth and produce local iron contamination. Regarding urban settlements, the majority of urban local bodies do not have underground drainage system and treatment facilities to collect and treat the municipal sewage (Fazi *et al.*, 2009), resulting in sewage directly entering the river system or agricultural fields without treatment.

Downstream of the river, fish kill due to pollution is frequent, affecting the livelihood of fishing families. Many fishermen have changed their traditional occupation of fishing to other jobs. Additional evidence on the composition of fish species also shows that some species have either reduced in numbers or become extinct over the years (Nagothu *et al.*, 2008), indeed, several local varieties of fish have become extinct. According to fishermen around 50% of the local breeds have disappeared or decreased in their population. During the field study of the STRIVER project, many fishermen expressed concern over the increasing use of chemicals or fertilizers in agriculture. Forty-seven percent of the fishermen indicated that water was polluted while 26% of them commented that fish kills had negative impacts in terms of health and reduced income and 9% reported a combination of health, reduced income and also witnessed fish kills (Nagothu *et al.*, 2008).

The Karnataka State Pollution Control Board (KSPCB), constituted by the Government of Karnataka in 1974 is the primary enforcing authority for the various environmental laws related to the Tungabhadra, including water pollution control. The Regional Pollution Control Boards are located at five districts in the Tungabhadra river basin carries out frequent river water quality monitoring. In addition, the KSPCB has initiated action plans to prevent river pollution in four towns under the National River Action Plan (NRAP), initiated by the National River Conservation Directorate of the Ministry of Environment & Forest, Government of India. However, derspite these regulatory measures, implementation has not been very effective in controlling pollution. The role of the KSPCB is however limited to the extent of sending constant reminders and issuing show–cause notices indicating institutional constraints. For instance, KSPCB is not able to ensure construction of sewage treatment plants in towns that are discharging sewage directly into the river. There is lack of understanding in addressing the problem from a holistic perspective. The authority of the KSPCB is limited the collection of samples and warning of polluters, no serious effort has therefore been made to understand the future impacts of pollution and ways of preventing it. An important note is that these problems cannot be pointed at regulatory authorities alone, as strong political will is also required to bring about the required changes.

Initiatives taken at different levels for the protection of riparian zones and the associated ecosystems include national parks and protected specific sites. However, there are no specific plans for protection in place. The Karnataka Biodiversity Board was set up in 2003 with a mission to target conservation, sustainable utilization and equitable benefit sharing (karnatakaforest.gov.in/English/index.html*)*. This board has set up over 575 biodiversity management committees in the local bodies of 15 districts. Among them, the bio-park in Shimoga is being developed as a model heritage site. Kuvempu biosphere reserve at Kuppalli is spread across 6866 acres, and 3230.50 acres has been declared a biosphere reserve. The fisheries department aims to protect endangered fish species and fish sanctuaries. Prohibited locations for fishing are located at four different locations along Tungabhadra. The Kudremukh National Park in the western region of Karnataka is one of the largest stretches of evergreen forests of low, mid- and high elevation along with a mosaic of shola-grasslands. Bhadra wildlife sanctuary is located in the tropical forests of the Western Ghats in Chikmagalur district, and covers an area of 492.46 $km^2$. It is home to a large variety of flora and fauna and it is regarded as one of the best wildlife sanctuaries in India. However, the existence of these natural parks and protected areas is not enough in itself to preserve the ecosystem habitat and biodiversity, especially as these areas represent a small fraction of the river basin and elsewhere the resources are managed in an unsustainable way (karnatakaforest.gov.in/English/index.html). Several government agencies have undertaken, with partial success, massive deforestation works to mitigate the damage to the forests and rivers due to mining. According to the government over 7.5 million Acacia, Eucalyptus and other exotic trees have been planted; these mono-culture forests are however not an adequate replacement for the diverse and unique natural habitats that have been already lost, Although forest and land laws specify that deforestation is not permitted, it still continues due to poor monitoring and implementation of the laws by government agencies (Manasi *et al.*, 2008).

### 8.3.2 Measures to ensure efficient and equitable use of water

Managing water as an economic and social good is important for achieving efficient and equitable use. Failure on the other hand to recognize the economic and social value of water has led to wasteful and environmentally damaging uses of the resource. In general, economic instruments involve the use of pricing and other market-based measures to provide incentives to consumers and all water users to use water carefully, efficiently and safely.

In TBSB, a rather inefficient practice exists, where users pay irregularly at different time intervals or on a flat basis as a consequence of the lack of, at most locations, a metering system. Water pricing does not consider the total cost of water distribution. The water users are encouraged by the local corporations to shift from paying on flat rate basis, to payment on a volumetric basis. Currently, pricing of water coverage exists for industries, drinking water and irrigation, with some exceptions. Local political priorities play a role to determine different price levels imposed to certain user categories or according to the average welfare level of the concerned society. The policies give irrigation a special consideration, since required fees are more or less 'political' fees to sustain the agricultural sector.

Traditional water uses are still in practice although there has been significant degradation of traditional water systems along the TBSB. Both water quantity and quality create the reason for conflicts regarding inter-sector allocation. Moreover, the regulation between the upstream and downstream part of the river basin and the consequent livelihoods of 10,000 fishing families create an additional reason for conflicts (Manasi *et al.*, 2008).

Several initiatives towards water conservation like rejuvenating traditional tanks, promoting rain water harvesting taken up by the government, are working towards strengthening the traditional water systems, particularly the irrigation tanks that are multifunctional, by deriving not only economical benefits but also societal and environmental benefits. The practice of linking water indirectly with land and the fact of *de facto* control by better-endowed persons only accentuates rural inequality and water use efficiency. The control of groundwater at the field level is governed by a *de facto* system of rights as determined by farm size, the depth and number of wells, pumping capacity, and the economic power (Saleth, 2005).

### 8.3.3 Institutional analysis and stakeholder involvement in water management

#### *8.3.3.1 Institutional structure*

India has a federal system of governance and water is not a subject under the jurisdiction of the Union Government but a State subject; although the Union can intervene in the case of disputes concerning inter-state rivers (Inter-state Water Disputes Act, 1956). The two riparian states in the TBSB, namely Karnataka and Andhra Pradesh, have their own set of different water governance regimes – water policies, laws, institutions and practices. The TBSB includes a wide variety of state and non-actors or interest groups with very little interaction between or amongst them. The state actors or stakeholders are the various government departments or autonomous corporations that deal with TBSB water matters at different levels – national, state, district and local levels.

Institutional integration demands close coordination between institutions whose decisions affect the quantity, quality or flow of water resources and dependent ecosystems. In Karnataka pollution control is separated institutionally from abstraction management in the TBSB. The State Pollution Control Boards (SPCB) of Karnataka and Andhra Pradesh states with their branches at the district level (Davanagere, Raichur, Shimoga, Bellary, Koppal and Chickmagalur) are responsible for water quality and monitoring and control of pollution, while water abstraction is controlled by the state Department of Water Resources. In Karnataka, the Department of Water Resources replaced the former Department of Irrigation (The department is technically responsible for water resources and their allocation across different uses and users including storage and conveyance of water, pricing, maintenance (Nagothu, 2008).

It is the SPCBs who have responsibility for the periodic collection of water samples from different stretches of the river and from various pollution hotspots, monitoring of water quality and imposing penalties for non-compliance with pollution control standards. Different state agencies look after sewerage and supply, drinking water, smaller water bodies, groundwater, water infrastructure and construction. There is however, a multiplicity of state institutions and departments looking after various aspects of water with very little interaction or coordination amongst them. A reluctance to share data across the two riparian states and even across different departments within a state is evident, with each department functioning in isolation and often leading to 'turf wars' between different departments and departmental fundamentalism. Water management as presented is not dealt with at the basin level, but at state level, and questions have been raised about the effectiveness of the SPCBs and their lack of contact with users. Traditionally, water management has been quite sectoral in India and it has had more emphasis on creating infrastructure (dams, irrigation channels) and other technological interventions. The departments have conflicting interests and often find it difficult to integrate their interests with a common goal.

Non-state actors or stakeholders also comprise a very wide variety of actors. First among these are the direct users of water (comprising farmers, fisher men, rural artisans who directly depend on the water resources for their livelihoods as well as the rural and urban population who use water for drinking, domestic use and for cattle and industries) along with organizations and interest groups like Water Users' Associations (WUAs), academic and research institutions engaged in the study of water and water use, and different types of civil society organizations that engage with water issues and water rights. The direct users are mostly unorganized, do not have access to data and are not involved in the production of knowledge and formulation of policies.

The Tungabhadra Board, which reports to the union-level Ministry of Water Resources, controls water allocations and power output from the Tungabhadra dam to the basin states, but it is not a licensing body (Figure 8.2). The board has been set up for implementing the provisions of the Krishna Water Dispute Tribunal (KWDT) award. The so-called Bachawat award was agreed in the mid 1990's and it stipulates a fixed scheme of water allocation in the Krishna basin, including the TBSB. The interstate TB board is conducting the operation of the Tungabhadra reservoir and the award is presently under revision. Currently, the tribunal has been reconstituted, as the earlier award has lapsed and renegotiation is pending.

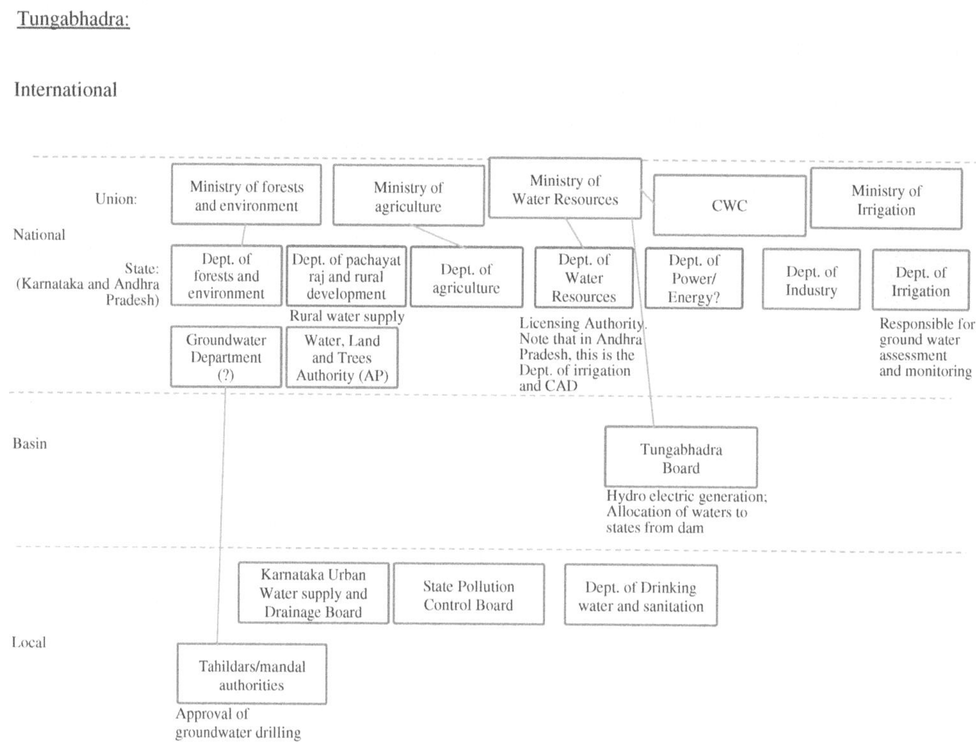


**Figure 8.2** Institutional structure in place for water management in the Tungabhadra sub-basin with indications of their roles where ever relevant.

### *8.3.3.2 Access to information for stakeholders*

Stakeholders do not have full access to information and data, and stakeholder consultation and dialogue is not mandatory, except for in relation to the Environmental Impact Assesssments (EIAs). In relation to an EIA, the department concerned is supposed to organize public hearings in which any stakeholder can put up his or her viewpoint. The experience so far however is that EIAs are not conducted properly. As the Tungabhadra does not have a basin-wide plan, interest groups have to rely on the detailed project reports (DPRs) of individual projects, which mainly outline project design, cost estimates, planned water use and designated cropping pattern. These documents do not usually contain governance issues, including stakeholder involvement. With respect to information on water resources assessment, some information is available within separate agencies, but for the moment it is not integrated and it is not available to the public. Records are also maintained on general human and health monitoring, but very often people do not get access to data and information and making data available to different stakeholders in an understandable form has been generally weak. Often there is no consistency in the data sets maintained by different agencies, which causes difficulties in negotiations concerning interstate rivers. A commonly agreed data set is an important condition for stakeholder involvement as well. The Karnataka State Water Policy (2002) states that a data information centre would be set up and data protocols developed to remedy the situation. The situation has also slightly improved with the Right to Information Act now in place in most of the states.

#### 8.3.3.3 Decision-making amongst stakeholders

There is a trend towards increasing involvement of users and stakeholders. However, there is no stakeholder involvement in the formal sense. Currently, water rights are governed by the National Water Policy (2002) and Irrigation Acts of the respective states. However, there is no clear legal framework specifying water rights, even though various acts have some basis for defining some form of such rights. The National Water Policy provides for the establishment of river basin organizations, and for the development and management of river basins or sub-basins, wherever necessary. Water policy documents and legislation, such as the Participatory Irrigation Management Act and 'The National Water Policy, 2002' call for a "participatory approach to water resources management". Similarly, in the Karnataka State Water Policy (2002) the word stakeholder appears in the context of participatory irrigation management and water users' associations to manage irrigation water. Though the National Water Policy (2002) and the Karnataka State Water Policy (Andhra Pradesh has yet to come out with its water policy) do mention stakeholder involvement, not much has been done in practice. In Andhra Pradesh the state government has passed a legislative act: *Andhra Pradesh Farmers' Management of Irrigation Systems Act* (1997), making participatory management of irrigation water part of the official policy of the government. It also established more than 10,000 WUAs immediately after this act. Although no such act exists in Karnataka, the state did encourage setting up WUAs on a voluntary basis and made suitable amendments to the Karnataka Irrigation Act (1965), in 2000 and 2002 (Nesheim *et al.*, 2008).

Through WUAs, Water Development Committees and self help groups, some opportunities exist for civil society participation with financial support provided for their attendance at meetings. Yet these opportunities play a very limited role, due to lack of resources, political will and priorities. Issues related to particular WUAs are discussed at a local level, but issues of policy etc., are discussed at the state capital. The public are not involved in problem identification and although public hearings are compulsory for clearing projects with environmental and displacement related impacts, the experience of many stakeholders and civil society organizations has an opinion is taken into account at some stages, but not formalized. The only area where involvement of the stakeholders is sought is in the area of irrigation water management as part of the sectoral reforms, thus almost all documents mention water user participation for irrigation water. In the drinking water sector efforts are being made to involve users in the management of drinking water schemes, both in the urban areas and rural areas. Micro-watershed development is a major program in the rural areas, funded by the Ministry of Rural Development, the Ministry of Agriculture and multi-lateral and bi-lateral agencies. Within the guidelines for community participation during program implementation, institutions such as village councils, a Watershed Development Committee and Self Help Groups have been set up and include the poor and women. Thus in all the three major sectors – irrigation, drinking water and watershed development – efforts are being made to bring in participation.

Despite this, decision-making is in general dominated by public officials particularly engineers, and multi-stakeholder fora are not in place. Various studies indicate that there is a large gap between the official/policy rhetoric and practice on the ground.

### 8.3.4 Capacity building

According to the Global Water Partnership, capacity building is the process of developing and strengthening the abilities of people, institutions and societies to perform functions, solve problems, and set and achieve objectives (www.gwp.org). In this way, a community equips itself to undertake the necessary functions of governance and service provision in a sustainable fashion. Capacity building can take place at different levels such as the central political institutions, central Governmental institutions and Departments, NGO's and Civil Society, and small businesses etc.

The Karnataka State Water Policy, 2002, does not say anything about improving IWRM competency. However, the policy does mention that efforts should be made to develop integrated management of water. Although capacity building is officially one of the main focus areas and part of many water policies and strategies in Tungabhadra basin, it is evident these official statements are seldom operationalized to any larger degree especially for the entire spectra of IWRM themes. Some examples of transparency (e.g., access to information), public participation, and information campaign brochures for the public, training programs for certain sectoral groups (especially for local communities, farmers) are however noted and usually initiated by NGOs. The relevance of this issue will vary according to the general state of knowledge/education and the extent to which the people interact with public officials.

The new Participatory Irrigation Management (PIM) policy in Karnataka also mentions that training of farmers and managers is essential for implementing PIM. In most water projects, capacity building is included as a part of the project activity, though it is not specifically mentioned as a capacity building exercise for improving the IWRM competence. A certain amount of budget is allocated for training, but this is not adequate to meet the entire training needs in the basin. During the farmers' interviews in Tungabhadra, it was observed that the majority of farmers expressed the desire to attend more training programs in water management. The Water and Land Management Institute (WALMI) located in

Hubli-Dharwad has been provided with a specific role to educate and train water managers, farmers and policy makers to get used to their new roles in the decentralized water management. Under the new PIM in the state of Karnataka (in which Tungabhadra is located) the government has made the training of farmers engaged in the Water User Co-operative Societies part of the new PIM program. Each year, policy makers, water managers and farmers undergo training programs designed to improve their skills in water management. The farmers' survey in Tungabhadra also revealed that, nearly half of the farmers in the basin have either experienced some kind of training program at WALMI or other agencies, or attended workshops or farmers meetings related to water management. In addition, the WUCS have regular meetings where farmers participate and meet up to discuss about water management problems. Officials from the Water Resources Department also participate in meetings that are organized at different levels. In addition, there are also public hearings organized by NGOs, like Jalaspandana, which educates farmers and managers on water related issues.

### 8.3.5 Transboundary issues

The management of transboundary waters has always been a difficult issue in which national legislation and international conventions meet each other within complex institutional contexts. While such agreements are made at the regional level, the actual policy changes, reforms and implementation of reforms need to happen at the national level. As mentioned earlier, the main issues at stake in the Tungabhadra River are the utilization of water for agricultural, urban and industrial uses, hydroelectric projects and pollution from industries, urban areas and agriculture. The Krishna Water Disputes Tribunal, from the early 1970s, is the main agreement for the use of water, and the so-called Tungabhadra Board oversees the distribution of the Tungabhadra reservoir water to the two States of Karnataka and Andhra Pradesh. Allocation of water for hydropower generation is also part of the agreement. The board is staffed by engineers from both states. On the 4th of August 2004, a New Krishna Water Dispute Tribunal was constituted. The main problem with implementation of these agreements seems to be that there are no practical procedural arrangements for solving controversies. Central government officials are actors and part of the tribunal, as well as respective state governments, who are also responsible for legal issues. In the case of short term and crisis management response to concrete issues there are many players, including local authorities and user organizations. Each state has an obligation to notify and consult each other on planned measures as per the Inter-State Water Dispute Act. However, while project proposals are accompanied by technical data, and there is a Central Water Commission which checks procedure, states tend to treat data very closely, and data exchange is not fully implemented, nor are there any statutory requirement for states to exchange data and information in general.

## 8.4 CONCLUSION

Integration is a concept that different stakeholders are realizing as an important step towards improved water management. One can see several initiatives in India and in TBSB that point towards IWRM, but still in the initial phase. In TBSB, it is fairly a long way to go. Strengthening science-policy linkage may improve the process. Besides, political will and awareness is necessary to realize the concept in practice. Strategies and basin plans are generally either recent innovations, or currently under preparation. The National Water Policy is positive towards IWRM, but it is the states that have to do that job. The state actors are the various government departments or autonomous corporations that deal with TBSB water matters at different levels – national, state, district and local levels. The absence of a legal framework that puts IWRM in place means that the dominant water user is more likely to be able to control and skew water resource management in its favor, a situation of high relevance as the differences in the standards of living are tremendous.

There is a need to incorporate a participatory approach in the management of the water resources. This should also include the participation of women. Such a participatory approach should involve not only the various governmental agencies but also the users and other stakeholders, in a decisive manner, in various aspects of planning, design, development and management of the water resources schemes. WUAs and the local bodies such as municipalities and *gram panchayats* (village councils) should particularly be involved in the operation, maintenance and management of water infrastructures/facilities at appropriate levels, with a view to eventually transferring the management of such facilities to the user groups/local bodies. Through user organizations, training and technical support communities and stakeholder participation must be empowered and public awareness created.

## 8.5 REFERENCES

Annual Report 2006–07, Tungabhadra Board (waterresources.kar.nic.in/)
Andhra Pradesh Farmers' Management of Irrigation Systems Act (1997) www.irrigation.ap.gov.in/iswr1.html
Karnataka State Water Policy (2002) Government of Karnataka (waterresources.kar.nic.in/)

Nagothu, U.S., Raju, K.V., Manasi, S., Latha, N. and Lenin Babu, K. (2008) Current Status and Future Possibilities to Improve the Livelihoods of Marginal Communities. STRIVER Report D9.5. http://kvina.niva.no/striver/Disseminationofresults/tabid/70/Default.aspx

Nagothu, U.S. (2008) Institutional Analysis of Participatory Irrigation Management in the Malaprabha River Basin, Karnataka, India, Bioforsk Report, Vol. 3 (120) 2008; ISBN-13 nr: 978-82-17-00417-2, Bioforsk, Ås, Norway.

National Water Policy (2002). Government of India (wrmin.nic.in/)

Manasi, S., Raju, K.V. Latha, N., Nagothu, U.S., Renault, L., Noemi, Vincente Serrano, S., Portela, M.M., Betamio de Almeida, A. and Machado, M. (2008) Competing Water Uses – Current Status and Issues in Tungabhadra and Tagus River Basins, STRIVER Deliverable 9.1, http://kvina.niva.no/striver/Disseminationofresults/tabid/70/Default.aspx

Nesheim, I., McNeill, D.P., Sekhar, N.U., Grizzetti, B., Allen, A.A., Barton, D., Begueria-Portuges, S., Berge, D., Bouraoui, F., Campbell, D., Deelstra, J., Garcia-Ruiz, J.M., Gooch, G.D., Joy, K., Lana-Renault, N., Machado, M., Manasi, S., Nhung, D.K., Paranjape, S., Portela, M.M., Rieu Clarke, A., Saravanan, V.S., Thaulow, H. and Vincente-Serrano, S. (2008) 'First IWRM Assessment report for the four Case Basins: Glomma, Tagus, Sesan and Tungabhadra, STRIVER Deliverable 5.1, http://kvina.niva.no/striver/Disseminationofresults/tabid/70/Default.aspx

Fazi, S. and Lo Porto, A. (eds.) (2008) Scientific report on Pollution source assessment, including source apportionment results, and pollution prevention strategies, STRIVER Deliverable 7.1, http://kvina.niva.no/striver/Disseminationofresults/tabid/70/Default.aspx

# Chapter 9

# Integrating tanks into the larger waterscape in the Tungabhadra

*S. Paranjape, K.J. Joy, S. Manasi, N. Latha and P. Mollinga*

## 9.1 INTRODUCTION

In India, an irrigation tank or tank may refer to an artificial reservoir of any size. In particular, when referring to old village tanks, it refers to earthen bounded reservoirs constructed across slopes by taking advantage of local depressions and mounds. Tanks are part of an ancient tradition of harvesting and preserving the local rainfall and water from streams and rivers for later use (Rainharvesting, 2010).

Tanks are an important part of the South Indian waterscape (for a definition of 'waterscape' see Molle, 2005). However, they are not well integrated into other larger systems. It is argued here that such an integration is important for IWRM strategies about tanks. The subsequent section describes the changing waterscapes in the Tungabhadra sub basin. It traces the changing historical contexts in which tanks have functioned, the changes in water profiles that have taken place and mark out the decline of tanks as a major component of this changing profile. The next section discusses the present situation of tanks in the sub-basin, especially in the upper reaches with the help of a study conducted under STRIVER. It concludes that tanks still play a significant part in the upper catchment and that they have multiple uses. It also points out that balance reservoir tanks have relatively less shortages and that tanks are used in conjunction with wells to serve a crop pattern dominated by a few water intensive crops. The next section describes the short term role that tanks could play in an IWRM system. This requires that their rehabilitation be carried out with appropriate priorities and that they be treated as multi-purpose community resources. It points out that they have the potential to be suitable instruments for dispersed equitable access if we are prepared to go beyond a simple participatory irrigation management approach. The next section describes a long term approach to this integration and the concluding section illustrates the measures that need to be taken in respect of both, large systems as well as tanks if this strategy is to be realized.

The study of tanks in the Tungabhadra basin referred to above was carried out as part of the STRIVER project (www.striver.no). We draw substantially on the policy brief on tanks prepared for STRIVER and on a special article written for SAWAS (Paranjape and Joy, 2010).

## 9.2 CHANGING WATERSCAPES IN THE TUNGABHADRA SUB-BASIN

### 9.2.1 Changing historical contexts

The traditional waterscapes in South India were dominated by tanks as Figure 9.1. Traditional waterscapes evolved slowly over centuries as societies adapted to their environments and tried to make the best use of the water regimes in their immediate ecosystem. Even though they show great variety, we may note three broad types that map on to three


Edited by Geoffrey D. Gooch, Alistair Rieu-Clarke and Per Stålnacke. ISBN: 9781843393252. Published by IWA Publishing, London, UK.

large regions of India. In mountainous regions North of the Ganga with the Himalayas snow fed perennial streams we find kuhls, diversion system, dominating the landscape In the Indo Gangetic plains we find the great river systems and their open access groundwater regimes dominating the waterscapes. In South India, we have tanks, often an interconnected cascade of tanks dominating the landscape. Almost every village had a tank and around them had also grown institutional systems that had evolved mechanisms of upkeep, maintenance and improvement. In the Tungabhadra sub-basin too (TBSB) it was tanks that dominated the traditional waterscape, though in the lower part of the basin that now falls in Andhra Pradesh, canal irrigation was also important. As we found in our studies, every village had a tank and often every hamlet had one too.

**Figure 9.1** Rice field irrigated by water tanks (Photograph: K.J. Joy).

A major change took place in the waterscapes in India with the advent of the British crown rule and modern construction technology. The British brought with them a new kind of waterscape – the dam and its command area based on modern science and technology, though they did not truly understand irrigation. The dams they built were rarely integrated with local sources. It introduced a system that was not as well integrated into the local social structures as the earlier waterscapes. This new system was based on science and technology and chose its locations according to its own set of rules. It aimed at supplying water for irrigation, sometimes as a drought relief measure, but mostly as a revenue maximizing device. These modern systems were introduced and the local systems were expected to adapt to them. Their governance structures and regulation systems were very different from those which marked the traditional waterscapes centered on the tanks.

The next major change in the waterscape took place after independence. First, though there was respect for the old village system and industries and also an attempt to nurture support and improve them, there was an increasing emphasis on modernization which is reflected in the often quoted remark, attributed to Pt. Jawaharlal Nehru the first Prime Minister of India, that dams and power stations constituted the temples of modern India. Both these temples have a role to play in changing the waterscapes in India. Large and small dams began to be built at a rapid pace. These dams had their own systems which as mentioned above were independent of the decentralized water sources and systems that existed in every village in India. They were seen as alternative, separate systems and the demands of administration and bureaucracy, approved and maintained their separateness. A parallel increase took place in power generation and after the sixties, when electricity began to rapidly enter rural areas, the electric motor-pump became a ubiquitous device, helping those in irrigation commands to trap percolation from the canal systems and in other areas to intercept and utilize, and now even mine groundwater.

### 9.2.2 The change in waterscape profiles

As has been noted in earlier chapters, water in India is not a federal subject but a state subject, and this has its effects on the shape that waterscapes take, as well as on the available information about water resources. Moreover, basin boundaries have no respect for administrative boundaries and their lack of coincidence is another data limitation that we have to cope with. The tables that follow attempt to show the direction change in water source profiles in the TBSB. We have chosen to present the data for the net irrigated area over a fairly long period, from the late 1950s and early 1960s to the middle of the first decade of the current century, the 2000s. The data are summary data for the districts in the two states that cover the basin – the Tungabhadra sub-basin includes portions of three districts, namely Cuddapah, Kurnool and Mahabubnagar in Andhra Pradesh and seven districts, namely, Bellary, Chikkamagalur, Davanagere, Haveri, Koppal, Raichur and Shimoga from Karnataka. As such the districts together account for more than the TBSB area, but the TBSB forms a sufficiently large proportion of their aggregate area so as to enable us to make assessments about broad trends within the basin. It may also be noted that the Andhra Pradesh portion of the TBSB is more or less the downstream portion, where the availability of canal irrigation is very high, while the Karnataka portion of the TBSB is the upstream portion of the basin where the availability of canal irrigation from large systems is not as high as in Andhra Pradesh.

Tables 9.1 and 9.2 below bring out the changing waterscape profile in Karnataka between the late 1950s and the middle of the 2000s. They show that net irrigated area more than doubled over the period. Most of this contribution is accounted for by a rise in canal irrigation, followed by bore-wells and a little by wells. In fact the contribution by tanks fell to less than half of what it was earlier. If we look at the proportionate contribution to net irrigated area, we see in the 1950s a waterscape dominated by tanks; they contributed more than half of the net irrigated area in the basin, followed by canals and other sources. Bore-wells were non-existent. At present, we see a waterscape completely dominated by canals which account for almost half of the net area irrigated, followed by bore-wells that account for more than one-fourth while tanks are marginalized; they contribute less that 10% of the net area irrigated. If we take into account the fact that a very large proportion of wells and bore-wells are dug in command areas to take advantage of canal seepage, the dominance by canal systems would stand out even more.

**Table 9.1** Net area irrigated by different sources in the Karnataka districts of the TBSB (ha).

| Year | Tanks | Canals | Bore-wells | Wells | Other sources | Total |
|---|---|---|---|---|---|---|
| 1957–58 | 157,156 | 56,934 | – | 18,279 | 62,155 | 294,525 |
| 2003–04 | 68,882 | 318,816 | 198,587 | 47,624 | 24,518 | 701,531 |
| %change | −56.17 | 459.97 | NA | 160.54 | −60.55 | 138.19 |

*Source: Directorate of Economics and statistics, Government of Karnataka*

**Table 9.2** Proportions of net area irrigated by different sources in the Karnataka districts of the TBSB (%).

| Year | Tanks | Canals | Bore-wells | Wells | Other sources | Total |
|---|---|---|---|---|---|---|
| 1957–58 | 53.4 | 19.3 | 0.00 | 6.2 | 21.1 | 100.0 |
| 2003–04 | 9.8 | 45.5 | 28.3 | 6.8 | 3.5 | 100.0 |

*Source: Directorate of Economics and statistics, Government of Karnataka*

The trends for the Andhra Pradesh portion of TBSB are similar. (See Tables 9.3 and 9.4 below.) There was a substantial, more than 50%, increase in net irrigated area during 1960–61 to 2004–05. The number of borewells registered an enormous increase, in fact, the number of borewells now in the Andhra portion which comprises three districts is even greater than the number of borewells in the Karnataka portion which comprises seven districts. Of course as mentioned earlier a lot of these borewells are in the canal commands. But their number in the Andhra portion is quite large. Here too the initial landscape has tanks as the most important source contributing about 40% of the net irrigated area in 1960–61, followed closely by canal systems which contribute more than 30%. The picture however, had radically changed by 2004–05. Canal irrigation contributes almost 25% to the net irrigated area, but the waterscape is now dominated by borewells which contribute almost 60%, while tanks' contribution has reduced sharply to a mere 3%.

**Table 9.3** Net area irrigated by different sources in the Andhra districts of the TBSB (ha).

| Year | Tanks | Canals | Borewells | Other wells | Other sources | Net area irrigated |
|---|---|---|---|---|---|---|
| 1960–61 | 108,829 | 84,794 | 0 | 73,273 | 8,717 | 275,613 |
| 2004–05 | 12,176 | 1,02,693 | 2,43,458 | 50,143 | 14,628 | 423,098 |
| % change | −88.81 | 21.11 | NA | −31.57 | 67.81 | 53.51 |

*Source: Directorate of Economics and Statistics, Government of Andhra Pradesh*

**Table 9.4** Proportions of net area irrigated by different sources in the Andhra districts of the TBSB (%).

| Year | Tanks | Canals | Borewells | Other wells | Other sources | Net area irrigated |
|---|---|---|---|---|---|---|
| 1960–61 | 39.5 | 30.8 | 0.0 | 26.6 | 3.2 | 100.0 |
| 2004–05 | 2.9 | 24.3 | 57.5 | 11.9 | 3.5 | 100.0 |

*Source: Directorate of Economics and Statistics, Government of Andhra Pradesh*

### 9.2.3 The decline of tanks

The continuous decline of tank systems in the post-independence period has been noted and discussed by a number of scholars (Von Oppen, 1980; Chiranjeevulu, 1992; Shankari, 1993; Agarwal, 1997; Vaidyanathan, 2001; Shah, 2002; Raju, 2003). The post independence decline has been ascribed to many factors. It is most commonly traced to a chain of events started by the takeover of community and zamindari tanks (private tanks) by the state. The state had its own rules and regulation in respect of small irrigation works that it applied to the tanks. As a consequence, the traditional institutions that governed the tanks began to break down and there was erosion of traditional arrangements in most tanks. The breakdown of traditional arrangements also led to confusion and deterioration of service. Traditional arrangements were designed to see that collection of charges was linked to maintenance, control over encroachments and general discipline. Their breakdown led to disruption of collection of water charges, lack of maintenance and increasing encroachments on tank beds and feeder channels. The decline also led to decrease in recharge of groundwater and increase in flash floods and overspills and reduced capacities.

Tanks were also made redundant because of environmental degradation in upstream catchments such as deforestation, overgrazing, soil erosion and siltation. In addition, changes in land use pattern particularly in the catchment zones of reservoirs, aggravated soil erosion and subsequent siltation in tank beds. With the extension of rural agricultural community beyond the traditional sections, neo-farmers are yet to acquire the proper agriculture and water management skills that came with traditional practices. However, those practices were often rooted in the availability of greater amounts of land per capita so that there was enough cushion for regenerative practices like fallowing, rotations and maintaining permanent cover on a substantial proportion of the landscape. With increasing pressure on land these practices rapidly disappeared.

However, in our opinion, there was another and larger factor that was also responsible for the decline of tanks. This was the belief that small irrigation sources were less efficient, much more costly and were a revenue burden for the state. This attitude can be traced back to the colonial period and was an attitude that has continued into the post independence period dominated by the view that the larger a project potentially the more reliable and more efficient it was. This has led to an attitude that sees the smaller and the larger systems as two separate competing entities rather than potentially different elements of an integrated system. Mollinga (2010) traces such a lack of integration to a state-village dichotomy where systems like dams and canal irrigation represent the state power and domain while tanks are more related to older systems and elites. Much of the lack of integration between tanks and large systems may also be related to the inability of the colonial as well as the post colonial state in India to integrate these two sub systems.

## 9.3 THE PRESENT SITUATION OF TANKS IN THE UPPER CATCHMENT OF THE TUNGABHADRA SUB-BASIN

During the STRIVER project, SOPPECOM with the help of two local organizations, Integrated Community Development Organisation (ICDO), Davangere and Vikasana, Shimoga took up a study of tanks in Shimoga and Davangere districts in the upper part of the Tungabhadra sub-basin. Five villages served by four tanks were taken up for study. Two types of tanks were included in the study. Two of the tanks were rainfed tanks and two of them were balancing tanks, that is, they served as buffer storages for the larger Bhadra irrigation system and therefore received periodic supplements which were stored in the tank. The net result was that these tanks received water from their own catchment as well as from the Bhadra system.

A house listing census was carried out in the 1655 households in the five villages. Of these a small stratified sample of 202 households was then selected for a more detailed household survey based on perception data. Householders were asked various questions and asked to provide reasons or comments and were also asked to rank them in terms of importance. The household data was then further discussed in focus groups to verify trends. Some of the interesting and relevant findings of the survey are given below.

### 9.3.1 Tanks still form an important irrigation source

One of the first findings was from the house listing. The house listing asked respondents how much land they cultivated and how much of it received irrigation and from which sources. The data show that of the total reported cultivated area of 1855 ha, 573 received water from the tank, a little above 30% of the cultivated area – this proportion varied from around 20 to almost 50% in the five surveyed villages (see Table 9.5) The proportion of land receiving water from tanks as a percentage of total irrigated area was about 60%; this proportion varied from a little above 40% to a little under 70% between the five surveyed villages (see Table 9.5).

**Table 9.5** Relative proportion of land irrigated by tanks in the surveyed villages.

| | Ha | % of cultivated land |
|---|---|---|
| Cultivated land | 1,885.65 | 100.0 |
| Land receiving irrigation from tank | 573.73 | 30.4 |
| Land receiving irrigation from other sources | 397.35 | 21.1 |

### 9.3.2 Tanks have multiple uses

The house listing form also included a simple question where the households were asked to indicate the different kinds of uses of tanks that the households employed. It clearly shows that tanks should not be regarded as irrigation sources alone. A very high number of households use the tanks for many other purposes.

Table 9.6 shows that the tank has a host of non-irrigation uses and a substantial number of households report these uses. In this table the use of tank as a source of drinking water is not covered, since it only covers those households that individually draw water directly from the tank for drinking. However, in almost all villages the tank is the basis of the drinking water supply for the village. What it shows clearly is that tanks cannot be considered to be simply irrigation sources. Indeed this applies to almost all reservoirs, ranging from the village tank right up to the large dams. This has major implications for the definition of water user and the organization of water users. Presently the state has chosen to treat as water users only those persons who hold land in the designated irrigation command of projects. This is too narrow a definition to adequately take account of actual water use and water uses with text: Two examples of multiple uses of tanks are given in Figure 9.2 and 9.3.

**Table 9.6** Multiple uses of tanks.

| Type of use | No. of HH | % HH |
|---|---|---|
| Drinking water (directly from tank) | 23 | 1.4 % |
| Washing clothes | 1,264 | 76.4 % |
| Bathing | 146 | 8.8 % |
| Drinking water for cattle | 673 | 40.7 % |
| Washing cattle | 673 | 40.7 % |
| Fishing | 22 | 1.3 % |
| Swimming | 229 | 13.9 % |
| Worship | 236 | 14.3 % |
| Tank bed cultivation | 18 | 1.1 % |
| Silt for farm land | 56 | 3.4 % |

*Total no. of HH = 1655*

Another use inadvertently left out of the house listing form is water used for sanitation. In the later sample household survey, households were asked whether they used water from the tank for sanitation and the findings show that according to season, between 65 and 82% of the households also use the tank as source of water for sanitation.

**Figure 9.2** Space for women to wash clothes (Photograph: K.J. Joy).

**Figure 9.3** Water source for cattle, Heere Mygalkere tank, Davanagere district. (Photograph: K.J. Joy).

### 9.3.3 Shortage of water

The reference year 2007–08 was not a bad year in terms of rainfall and water availability and not many households reported water shortages. However, almost everyone complained of recurrent shortages. When asked to assess how many of the last ten years were shortage years, the average of the reported value was 3.2 years out of ten.

Households were also asked their opinion of the reasons for water shortage and were asked to rank them. The top five reasons that ranked the highest according a simplified rank index were:

(1) Lack of repair and maintenance of tanks
(2) Excess water drawn by farmers at head reach
(3) Encroachments on tank bed have reduced storage
(4) Too many people grow water intensive crops
(5) There is less rainfall

Interestingly, excess water uses as well as encroachments are also quite important perceived factors. This has implications for how to go about the tank rehabilitation programs.

### 9.3.4 Wells and conjunctive use

There is a preponderance of bore wells (total of 52 bore wells owned by the sample) as against dug wells (only 6). Average pump rating was 5 hp and hrs of operation per week varied from 6 in the monsoon to 13 in winter and 21 in summer. Most of the wells go dry in February. However, almost all of the wells are recharged when the tank channels are flowing. This is again important to note: these wells will statistically be counted as separate irrigation sources and area irrigated by them will appear as well irrigated area. However, many of the uses of water from these wells will come from tank water replenishing them and therefore a substantial proportion of the irrigation they provide is indirectly tank irrigation. Unfortunately, there is no simple way present statistics can segregate the two components. What is important is that almost universally now in tank and canal commands farmers routinely use wells to capture recharge from irrigation and such use is now common.

Households were also asked about the major problems they faced regarding wells and the top five were as follows:

(1) Electricity is available only for short while
(2) Electricity comes in the night
(3) Well yield is too low
(4) Electricity charge is too high
(5) Wells go dry

As may be seen, electricity is closely implicated in well-related problems.

### 9.3.5 Crop preferences

Crop preferences as revealed by ranking crop choices made by households were as follows:

(1) Paddy
(2) Maize
(3) Areca nut
(4) Sugarcane
(5) Sunflower

Of these, paddy, areca nut and sugarcane are all water intensive crops and it is these that are at present accounting for a very large proportion of irrigation water use. A major thrust on water saving technologies in these crops could thus yield substantial water savings. However, there is also a need to link such savings with pooling and providing them to those areas and farmers who at present are deprived of water. If this is not done, the water saved will simply contribute to expanding the area of water intensive crops grown by the resource rich.

In conclusion, it may be said, that although tanks are in decline, they continue to play an important role in the village waterscape. They served between 20% and 45% of the cultivated area, which formed between 40 to 70% of the irrigated area in the surveyed villages. A substantial number of villagers depend on it for many of their water needs. They are not to be seen only as irrigation sources but are multifunctional entities that serve a variety of interconnected needs including the spiritual and cultural needs built around water. The findings also point to the need to go beyond simple desilting in order to rehabilitate tanks. The issue is how can his decline of tank systems be arrested and how can they be meaningfully integrated into the larger waterscape?

## 9.4 TANKS CAN PLAY A VITAL ROLE IN EQUITABLE AND SUSTAINABLE IWRM

The decline of tanks hides their actual potential. There is a need to study what their role could be if they are properly rehabilitated and integrated into the larger waterscape. In one such study, Raju (2003) estimates the potential of tank irrigation in Karnataka to be as high as 690,000 ha which is comparable to the 743,383 ha irrigated by large canal systems in Karnataka in 2003–04. Moreover, there is also increasing evidence that the notion that tank irrigation is not efficient because of its scale is not borne out by the available information. For example, Raju (2003) shows that the data according to the data from the Ministry of Water Resources itself, the cost of creating irrigation potential for one hectare during the Eighth Plan through large and medium irrigation projects was Rs 98,495 as against just Rs 10,051 for small irrigation projects. This indicates that tanks can possibly play an important role in livelihood assurance and poverty alleviation. For this to happen and their full potential to be realized there are a number of changes that have to be brought about in both, how we think about them and how we deal with them in practice, giving due attention their advantages and limitations.

### 9.4.1 Prioritizing rehabilitation

It goes without saying that tank rehabilitation is a necessary component of any strategy that attempts to exploit the potential that village tanks represent. However, it is important that both the village communities as well as the state move away from the viewpoint that sees rehabilitation largely in terms of de-silting. Village communities see it as an important employment generation activity with state support. It cannot be emphasized enough that de-silting creates additional storage at great cost and that its benefits are liable to be nullified if other concomitant measures like catchment protection, etc., are absent.

Our survey corroborated this viewpoint and also gave us indications about prioritizing rehabilitation work. Accordingly we would suggest that in rehabilitation measures, the first priority should be to clear encroachments, first, on to the feeder channels and supply channels as well as overflow weirs if present, and secondly, on to the tank bed. Next should be cleaning of gates and channels, repair of walls and embankments, provisions for facilities and sites for washing clothes, collecting drinking water, separate sections for animals for drinking and washing, and proper drainage from these sites. Prevention of de-siltation through filters and catchment treatment comes next, and de-siltation comes at the end. It is important to build source protection into tank rehabilitation.

### 9.4.2 Multi-purpose community resources

Tanks have to be looked upon as multifunctional community resources, micro waterscapes embedded in their ecosystem and socio-economic-cultural contexts. It is important to move away from the conventional Water Users Association (WUA) approach that sees water mainly as an irrigation resource. From an IWRM perspective tanks have both direct and indirect uses: (a) direct uses – irrigation, drinking water, water for domestic use and sanitation, water for bathing, drinking water for cattle, water for washing and bathing cattle, washing clothes, fishing, recreation, worship, silt and seasonal tank bed cultivation, etc., and (b) indirect uses – basically ecosystem services that regenerate the environment, recharge and replenish ecosystem resources and potentials.

Looking at tanks as irrigation sources leads to identifying water users as those landowners who use water for irrigating their lands. This is actually the change that was brought about by the British, especially in their revenue maximizing behavior. Earlier forms often had an element of forced labor, and pre-colonial rulers also extracted revenue, but by an large the governance structure treated the tanks and indeed water resources in general, especially surface water sources, as community resources and had a much more integrated approach to them. Aesthetic, spiritual and environmental concerns were as important to them as irrigation and other direct uses of the water. Of course, they could rise above the inequalities and oppression that characterized traditional societies. It may not be possible to simply go back to those forms, but it should be possible to learn from them and devise community oriented forms that are equitable and sustainable. For example, traditional viewpoints tend to concentrate purely on local and community forms as autonomous units. It is important to take account of the administrative as well as the political importance of governance form at higher scales and come up with nested institutions that are community oriented.

### 9.4.3 Potentially more dispersed, more equitable IWRM instruments

Tanks systems are much more dispersed than the large centralized commands organized almost exclusively around irrigation. This dispersal has its own importance. Within the framework of an equitable water right independent of actual landholding along the lines advocated by Pani Panchayat and the South Maharashtra Equitable Water Movement, this dispersal has the potential to redistribute access in a regionally equitable manner and reach poorer farmers. However, this is not possible without equity forming an important founding principle of water resources governance, and this involves great political difficulties, given the concentration of economic and political power in the hands of local elites in the case of tanks and rich peasants and capitalist farmers in the large command areas.

For this reason, IWRM does not only require tank level IWRM institutions alone but also requires nested institutions at various scales. At the bottom most level the tanks serve as appropriate units of organization for IWRM because they combine local hydrological and associated ecosystem uses and users and bring together the many uses of water in a local ecosystem. Tanks therefore have great potential of taking us towards IWRM and widely dispersed livelihood assurance, provided we have a suitable policy aimed at realizing this potential. To this end the following suggestions emerge from the study.

### 9.4.4 Need to go beyond participatory irrigation management

The first step in this direction is to shift from Participatory Irrigation management (PIM) to participatory management of ecosystems and water. However, doing this is not a simple matter. In Karnataka, all tanks with commands of less than

4 ha are owned by the Gram Panchayat, between 4 to 20 ha by the Zilla Panchayat, between 20 and 200 by the Minor Irrigation Department and above 200 ha by the Major Irrigation Department. In Andhra Pradesh, tanks with a command/ irrigated area greater than 400 ha are completely with the Irrigation Department, Panchayat raj institutions maintain tanks below 40 ha in Telangana and below 80 ha in the rest of Andhra Pradesh, with joint charge for those in between.

Things have become more complicated after the advent of PIM. Andhra has gone much farther than Karnataka in respect of PIM experience and legislation. The PIM discourse is conducted in terms of 'water users', but the operational part always reduces water users to irrigators. In fact, PIM statutes in most states designate land owners in designated commands as water users and then proceeds to frame rules and regulations around this definition. This may be appropriate for large and medium systems that came up explicitly as irrigation systems, though in actual practice water use from canals is also quite varied, but it is surely not adequate for tanks which have come up and have been managed as multifunctional community sources.

This implies that the definition of water users and membership of WUAs must not be restricted to landowners in designated commands but must embrace the entire community that inhabits and utilities the immediate ecosystem that the tank comprises. Also, governance institutions should be based on a representation from the irrigation users as well as the community that effectively uses the tank which may be a habitat, ward, village, or group of villages according to the size and situation of the tank. This is a complex but necessary task if we are to move towards IWRM.

## 9.5 INTEGRATING TANKS INTO THE LARGER WATERSCAPE: MAKING INTEGRATION A POSITIVE SUM GAME

Rehabilitation measures described above constitute a road-map for the rehabilitation of tank systems. Though an important and necessary step, it still does not touch upon the relationship of tanks and the larger waterscapes and how that relationship. For that we need to approach the decline of tanks from a slightly different perspective. The present and largely popular view is that tank rehabilitation is seen to be the solution meant to arrest the decline of tanks. The argument implies that tank systems declined because their condition and the services they offered deteriorated. However, it is possible that if the deterioration of tank systems and the services they offered were not a cause of the decline but rather an effect of that decline. In that case it would be important to identify the other cause of that decline and try to deal with that cause.

In that context it may be interesting to see what actually happens that when canal irrigation enters an area. The farmers who own land in the canal command gravitate towards it and drop out of tank institutions. They think that water from the larger system will be, and it often is, more reliable than that of tanks. The overall tendency to neglect tanks and tank institutions intensifies manifold and tanks decline much more rapidly in canal served areas than they do in areas outside canal commands. While canal irrigation does add to irrigation in an area pre-existing irrigation is often quite substantial.

### 9.5.1 Tanks that receive water from the larger system

The first step in this direction is to recognize that tanks have a limitation; that while they do very well in good rainfall years, they do not perform well in bad rainfall years. In contrast most large systems show less variation and tend to perform better in bad years than small sources. This is often the reason why the engineer tends to favor large systems over small systems. However, if we think of tanks and larger systems, not as alternatives of one over the other, but ask what does this say about the relationship between the two, we get a different kind of answer. The answer is that stabilizing supplements from larger systems will strengthen the tank system.

Tamil Nadu had a number of so called system tanks – tanks which receive regulated supplements from larger systems. Studies like Vaidyanathan (2001) show that the performance of system tanks in Tamil Nadu is better than other rain-fed tanks. Karnataka does not have system tanks in the way that Tamil Nadu has. But it has something similar. It has the practice of using some of the tanks as so called balancing reservoirs, as buffer storages to hold water from the larger systems and extending canal reach. Our study of tanks in the Tungabhadra basin also shows that farmers in villages served by rain-fed tanks reported between 3 and 5 years out of every 10 years as shortage years, those from villages served by balance reservoirs reported less than 3 years of shortage. This improvement in spite of these tanks not being system tanks in the true sense of the word indicates that tanks that receive supplements from larger systems would be much more reliable and perform much better.

Tank supplements play a number of roles that are important from an IWRM perspective. Firstly, as pointed out above, since water from larger systems has a greater dependability of supply, such supplements can go a long way towards stabilizing tank supplies for various uses, not just irrigation. With growing awareness of environmental issues, large systems have come under criticism for creating ecosystem 'islands' within the larger ecosystems. Since tank commands are much more dispersed, providing tanks with stabilizing supplements from larger systems these supplements provides greater dispersal of water and helps mitigate the 'island' effect they are often criticized for. Thirdly, if availability of such supplements is made dependent on augmentation and protection of local resource and good performance, it can become an incentive and instrument for ecosystem improvement and sustainable productivity enhancement. Such integration of tanks into larger systems will go a long way towards dispersal of water access and poverty alleviation.

### 9.5.2 Breaking the conceptual barrier: integrating irrigated and rain-fed farming

However, this is unlikely to take place unless we break down the conceptual and institutional barriers between irrigated and rainfed farming. At present radically different policies are adopted for them and are implemented by different departments. Similarly, there is a divide between tanks (seen as 'minor' irrigation systems) and medium and major irrigation systems. But both are facing some sort of crisis. Rain-fed farming, seen as farming without any applied water, is rapidly declining, and farmers are abandoning it and crowding into the cities in search of gainful employment. At the same time, centralized canal irrigation systems are becoming more and more expensive, and are having to overextend themselves due to increasing demand for irrigation since irrigated farming is seen as the only worthwhile farming. The more centralized and larger they are becoming, the more problems they are having in assuring timely and sufficient deliveries to farmers' fields.

However, there are developments outside the canal systems that offer clues as to how integration may take place. First, there is the learning from traditional systems. Traditional systems were always considered rainfed systems. However, most traditional systems had a mix of irrigated and non-irrigated farming. Farmers usually practiced both and there was a variation across them both in terms of crop and plots of land. So, there is a recognition that farmers operated an integrated irrigated non-irrigated system. Moreover, rainfed farming practices that operated within these systems often had elements of risk minimizing and productivity enhancement that were based on having extra land area to trade off. Given the land constraints imposed by a growing population size, it is now being conceded that some irrigation is necessary for ensuring sustainable livelihoods. It is essential to see that this does not mean that all the farmers land is irrigated. The growth and spread of well irrigation in rain-fed areas is now leading to the development of a composite system that would benefit greatly from dispersal of access through strengthening and expansion of the tank system.

Traditional systems always had elements of such a flexible integration of irrigated and non-irrigated areas. It never managed and did not think in terms of fixed designated blocks of command areas. It was the village, the farmers, who ran it to serve their interest and had within it the inbuilt mechanisms to respond to need. Systems were laid out, at least in their inception stages to see that everyone got some benefit. The phad system in Maharashtra combined *phads* as irrigated contiguous blocks even while ensuring that everyone had some land within the *phad*. The systems were strictly equitable, they were after all inscribed within highly iniquitous social and economic systems and had to work within them. What is important is that if we look at them not as systems to be taken over but as systems that have definite lessons for us to learn and apply – they can help us deal with our own situation, in our own ways.

### 9.5.3 From command areas and irrigators to service areas and water user communities

Firstly, irrigation from this point of view is not a matter of providing irrigation water to designated pieces of land, but is a matter of providing irrigation service to farmers and communities. Secondly, water sources are not to be reduced to irrigation sources, but are treated as sources of fulfilling the whole gamut of water uses that a community has. These aspects were inbuilt into traditional systems, and modern systems have broken up these composite systems into individual, virtually independent systems. In that sense, we should look at traditional systems as practicing IWRM systems that modernization has fragmented and the issue is now that of their *re*-integration into a *new* IWRM system.

Managing such a system is different from managing canal irrigation systems in vogue today. Managing irrigation systems is seen as managing a block, and most often a centralized block of land in which all the land is potentially irrigated. In reality, we shall have to deal with managing a dispersed and changing irrigated area within a larger designated *service area*, rather than a given block of irrigated command area. The irrigated area could also be flexible, especially if we take into account that crop rotations and rotation of irrigated area is an important aspect of environmentally friendly sustainable practices and is also important for avoiding the irrigated island effect that centralized irrigation systems have. As an illustrative example, instead of a fixed designated command area of 500 ha that may be supplied with 1 Mm3 (million cubic meter) of water, such a system may have to deal with a changing pattern of say providing 1 Mm3 of water to between 400 to 800 ha within a service area 2500 ha, where farmers will combine rainfed and irrigated agriculture. The latter in fact is what is closer to what farmers already do.

The service area approach also implies that what is being provided is not irrigation alone but a host of water services, including for example, drinking water, water for sanitation and hygiene, water for cattle and other livestock, water for washing clothes, possibly fishing or fish culture, and water for recreation and religious purposes. Many of these services, especially those tied to cultural or spiritual needs are not simply the provision of water, but waterscapes, that is, particular combinations and placements of water and landscapes. It implies a radical move away from water users as owner-irrigators to water user communities. As water users in the widened sense they form a complex agglomeration that takes them from being a so-called single interest group to a community composed of intersecting and overlapping stakeholder groups. Indeed, even though a very large proportion of traditional tank water was used for irrigation, there was never any question that they were managed by the communities they served.

This is another point where tanks could have an important role to play. Community bonds do not form very easily and it is useful if there are natural-social affinities that can serve to bring about this bond among water users. Here it would be important to organize the lower rung of the irrigation system in such a way that the smallest units also coincide as much as is possible with traditional units, either in the form of a village or rather hamlet or group of hamlets or in the form of earlier tanks. This then would form the most appropriate lowermost unit for a nested IWRM system of water governance.

### 9.5.4 Grounding integration: minimum water assurance

As pointed out earlier, the integration of rain-fed and irrigated farming implies that farmers will normally have both irrigated and rain-fed land and that they will manage as a farming system that is composed of both elements and the task of irrigation providers is to provide irrigation facilities to such an integrated system. Irrigation provision must then take due note of how the integrated system works and its demands. If this is taken as a starting premise farmers can then rotate crops and irrigation in a manner that will avoid the islanding effect of centralized command systems and minimize the environmental ill effects of repeated and continuous mono-cropping.

Command areas were earlier designated by working out how much contiguous area could be irrigated by a designated depth of water and a hypothetical crop pattern. The new service areas are oriented towards ensuring livelihood security/ for as many *farmers* as possible, and the area they can serve is then derived from there. Planning the new service area then is like a micro planning exercise, in which the attempt is to identify livelihood needs and water requirements and gaps. The important thing here is the change from land to people as unit and from the introduction of livelihood assurance (which include all basic needs and not just irrigation) as the principle to be utilize in determining the service. Obviously we also need some guidelines on how much water could be then provided on the one hand to the more enterprising farmers for profitable activities and on the other, for the needs of environmental integrity and sustainability of the water source itself.

### 9.5.5 'Melons on a vine'

So far we have only described, in an illustrative manner, the lowest rungs of the IWRM system we have in mind. And, to that extent, many people would agree, but would also maintain that while it is appropriate for smaller systems and traditional tanks how do we then extend it to the medium and major systems, which may be at the scale of hundreds and thousands of villages? However, the problem is not all that intractable if we break it down into modular units. And ironically, a rudimentary form of that type of organization is also prefigured by the clues towards integration that are emerging from the larger systems themselves that we have pointed out earlier.

The clue to the organization of such an IWRM system on larger scale is afforded by the long-prevalent system tanks in Tamil Nadu and the increasing use of tanks as 'balancing' reservoirs in other areas. Let us imagine a system in which we have water communities with service area planned around local storages, each of them forming a modular unit. Imagine then that the medium or major systems are then organized in manner in which they deliver water, not to the end user but to these modular systems.

The whole system looks like what the Chinese have called 'melons on a vine', the melons being the local water systems and the vines the larger systems supplying the melons with the 'nutrients' they need! The system would be extended by the melons growing bigger or by adding new melons. These local systems then act as rainfed tanks in their own right but also as distribution mechanisms for the larger systems. In fact, if tank rehabilitation programmes are planned in such a way as to integrate them into the nearest larger system and the larger systems in turn are reorganized in such a manner wherever possible, we shall be going a long way towards an equitable and sustainable IWRM.

## 9.6 CONCLUSION

The transition to such a tank system is not a small task and will have to be taken up from both ends: in restructuring both the existing large and small systems and their further development. This is a big exercise. What follows are a few illustrative suggestions.

### 9.6.1 Approach towards local water systems

Three key areas are important to emphasis in this context. First, commence with rapid local water system development with watershed development as a starting point, then extend such development to all areas including irrigated areas, provided they adopt this type of integration. This will be important in converting present command areas onto melons on a vine.

Second, adopt Low External Input Sustainable Agriculture (LEISA) systems with rain-fed as well as irrigated components with limited but assured water application. The irrigated areas could comprise of small plots with intensive cultivation and the rain-fed component could be biomass production from rain-fed horti- and silvi-culture integrated into farming systems.

Third, extensive local buffer storages (e.g., farm ponds) for protective and limited irrigation (5% farm pond program) should be provided.

### 9.6.2 Approach towards larger water systems

All incomplete and new dams and storages should be restructured on the basis of the 'melons on a vine' concept, pool command area development funds with watershed funds to rehabilitate and develop local water sources through watershed development as a starting point.

Those concerned should restructure as much of the existing command area as possible on the same lines by rehabilitating tanks and local storages within and around the commands and integrating them into the system.

Support should also be provided for water saving and LEISA techniques for the major water consuming crops in canal areas with the condition that saved water be available to non-canal areas.

In addition, the quantum of water supply to a local system should be tied to the degree of development of local resource. Typically, it could be made proportional to the degree to which the local system realizes the sustainable water resource potential of that area. Since this ties supply of external water to the degree of development of potential of local sources, it helps avoid the sort of neglect and decline of local sources that follows the entry of canal systems into an area.

To summarize, it is our belief that tank rehabilitation without the long view in mind is insufficient in respect of moving towards a truly IWRM system that integrates the large and the small. Without the long view, tank revival will certainly take place, but will always remain threatened because it will lack the stabilizing integration into larger systems that it sorely needs. On the other hand, the larger systems will continue to be overextended, over centralized and will suffer from lack of sufficient buffer capacity. Both large and small systems will suffer. Together, they have the potential to form a stable IWRM system that will go a long way towards providing sustainable livelihoods for the rural areas in India.

## 9.7 REFERENCES

Agarwal, A. and Sunita, N. (1997) Dying wisdom: Rise, fall and potential of India's traditional water harvesting systems. Centre for Science and Environment, New Delhi.

Chiranjeevulu, P. (1992) Tank irrigation and agricultural development, Kanishka Pub. House. Kajisa, K. Palanisami, K. and Sakurai, T. (2006) Declines in the collective management of tank irrigation and their impact on income distribution and poverty in Tamil Nadu. *India FASID Discussion Paper Series on International Development Strategies* No.2006-10-005.

Molle, F. (2005) Elements for a political ecology of river basins development: The case of the Chao Phraya river basin, Thailand. Paper presented to the 4th Conference of the International Water History Association, December 2005, Paris, http://www.iwmi.cgiar.org/assessment/files_new/research_projects/River_Basin_Development_and_Management/Elements_Chao%20Phraya%20IWHA%20Paris2_Molle.pdf last accessed 2010-04-24.

Mollinga, P.P. (2010) The material conditions of a polarised discourse clamours and silences in critical analysis of agricultural water use in India. *Journal of Agrarian Change* **10**(3), 414–436.

Paranjape, S. and Joy, K.J. (2010) Tanks in the Tungabhadra sub-basin and an IWRM strategy for tanks: The long and the short view. *South Asian Water Studies* **1**(3).

Rainharvesting (2010) Water Tanks, http://www.rainwaterharvesting.org/Rural/Dhan.htm

Raju, K.V. (2002) Tank rejuvenation in Karnataka: why it should be a community based approach? IWMITATA Water Policy Research Program, Annual Partner's Meeting.

Raju, K.V., Karanth, G.K., Bhende, M.J., Rajasekar, D. and Gayathridevi, K.G. (2003) Rejuvenating Tanks: A socio- ecological approach. Books for Change, Bangalore.

Shah Tushaar, R., Seenivasan, C.R., Shanmugam and Vasamalai, M.P. (2002) Sustaining Tamil Nadu's tanks: A review of issues and approaches. In *Institutionalizing Common Pool Resources*. Dinesh K. Marothis (ed.), Concept Publishing Company, New Delhi.

Shankari, U. and Shah, E. (1993) Water management traditions in India. PPST Foundation, Chennai.

Vaidyanathan, A. (ed.) (2001) Tanks of South India. Centre for Science and Environment, New Delhi.

Von Oppen, M. and Subba Rao, K.V. (1980) Tank Irrigation in semi-arid tropical India. Economics Program Progress Report, ICRISAT, Hyderabad.

# Chapter 10

# Water quality assessment and water pollution modeling in the Tungabhadra River basin

*A. Lo Porto, A.M. De Girolamo, A. Gosain and L.J. Barkved*

## 10.1 INTRODUCTION

The loss of nutrients and sediment from land originates from both diffuse, i.e., non-point sources, and point sources. The combination of source type, source emissions and transport processes will determine the overall nutrient loss to waterways. In agriculturally dominated river basins, diffuse losses generally dominate the inputs of N and P into streams.

The management of water pollution from diffuse sources can be seen as a rather complex challenge because:

- non-point source pollution originates as a result of landscape process acting on large areas;
- pollutants reach water bodies after interacting with the soil system where complex chemical-physical transformations happen;
- hydrological processes play an important role in the transport of pollutants; and
- heavy precipitation events may lead to dynamic and unpredictable transport of pollutants, e.g., during monsoon rain events.

Tools for the quantification of diffuse sources of nutrients in river basins are strongly needed. Managers and policy makers increasingly rely on outputs from river basin models when evaluating management actions and environmental changes. In Europe, the EU Water Framework Directive (WFD) requires the member states to develop river basin management plans for each river basin. Hence, responsible authorities will need tools to assess alternative management options.

The river basin planning process is carried out through a series of steps:

(1) characterize existing conditions;
(2) identify and prioritize problems;
(3) define environmental and management objectives;
(4) develop scenarios and protection or remediation strategies; and
(5) implement and adapt selected actions.

Scenarios, based on the perception and the interests of the stakeholders in particular, can indicate possible future directions for development. The consequences of these scenarios on water quality and quantity can be evaluated by river basin modeling. This chapter describes the results of the application of such a procedure within the Tungabhadra River basin in India.


Edited by Geoffrey D. Gooch, Alistair Rieu-Clarke and Per Stålnacke. ISBN: 9781843393252. Published by IWA Publishing, London, UK.

## 10.2 DESCRIPTION OF THE STUDY AREA

The Tungabhadra River is a tributary of the Krishna, which finally flows into the Bay of Bengal. Tungabhadra is the largest tributary of the river Krishna, contributing an annual discharge of 14 700 million $m^3$ at its confluence point to the main river. The river has a several dams built across it, the largest ones are the Gajanur and the Tungabhadra dams. It has a drainage area of 71 417 $km^2$ out of which 57 671 $km^2$ lies in Karnataka State, after flowing for a distance of 293 km and the remaining downstream part in the state of Andhra Pradesh (AP).

The sub-basin is mostly rainfed, dominated by red soils, with an average annual rainfall of 1200 mm. The major crops grown are paddy, jowar (sorghum), sugarcane, cotton and ragi (millets). The Tungabhadra reservoir has been constantly losing its water storage capacity over the decades much to the concern of the governments. About 50 years ago the capacity of the reservoir was 3.6 thousand billion cubic meters which now have been reduced by around 25% due to siltation. The amount of rainfall has also decreased in the past few years and as the reservoir does not get filled up, water is released for only one crop now. Furthermore, conflicts arise between Karnataka and AP because of the increasing storage capacity and water use in the upstream part of the basin so that in lean years no or very little water reaches AP. The river basin includes a number of industrial activities in small and large plants and a wide range of commercial agricultural activities as irrigated agriculture has rapidly taken over areas under rainfed farming. Further information about the Tungabhadra River basin can be found in Chapter 8.

Traditional farming systems, which propagated the use of practices like green manuring and organic recycling, have declined. Even though the shift from traditional sustainable farming system to the intensive 'Green Revolution' agriculture brought about a marked increase in gross food production, it also resulted in soil depletion, reduction in land productivity in other parts of command area and the spread of monocultural crops. Encroachment on public land for cultivation is a common feature and has resulted in more land being brought under cultivation at the expense of tree cover. Excessive irrigation for cash crops, especially plantations, has also led to water stress in the region. The mining of iron ore at Kudremukh and manganese ore in Sandur in the upper river basins of Tunga and Bhadra has seriously affected the environmental status of the basin. The severe soil erosion and silting of several small reservoirs, traditional tanks and the Tungabhadra reservoir, have affected the irrigation needs. Forest degradation and mining has resulted in flash floods, increased erosion and pollution of the rivers and adjacent land.

The principal sources of pollution in the Tungabhadra River basin are: (a) run-off from agricultural fields; (b) industrial effluents; (c) sewage from urban settlements; (d) mining activities; and (e) over exploitation of ground water. Domestic and industrial pollution, combined with deforestation, use of pesticides and fertilizers have affected water quality extensively making water unfit for drinking. Agricultural activities have resulted in increased salinity, alkalinity and water logging problems in the command area. The use of fertilizers has also affected ground water quality in the basin. Nitrate is the major contaminant in ground water samples, reaching level of up to 1183 mg/l (DMG, 2004).

Tungabhadra River is the major source of water for several existing or planned large-scale industries located on its bank. Around several small-scale industries operate in the river basin. The major types of industries are iron, steel, paper, paper pulp, chemicals and sugar.

The Tungabhadra River basin has a large population. In the Karnataka part of the basin, around 11 million persons are resident, with an average population density of 223 persons/sq km. In urban areas the population density is around 10 times higher and a major contribution to pollution, as the majority of settlements do not have underground drainage systems and treatment facilities to collect and treat the municipal sewage. Sewage therefore directly enters the river system and irrigation canal system without treatment.

## 10.3 APPROACH TO HYDROLOGICAL MODELLING

River basin models are tools to represent the hydrological and bio-geochemical 'functional' behavior of a river basin system. In an IWRM context, these tools can provide – at least potentially – a knowledge base and/or scientific basis for comparison of present and future management options in the river basin concerned. Therefore, models might help raise awareness about the implications of policies and/or management strategies, as well as the effects of climate and land use change.

It should be noted that the models differ according to the processes included, the temporal and spatial resolution of data required, and type of results they can provide. It should also be stressed that the choice of modeling tool largely depends on the specific objectives of the study.

Models need to be parametrized, calibrated and validated for the specific local conditions being studied. These steps require general physical and management data information on the river basin and the availability of adequate and more specific monitoring data on water quality and quantity. In some cases, there is also a clear need to apply models where data are scarce. However, the availability of adequate data is a crucial prerequisite for model applications, as the accuracy of the predictions and the quality of the assessment are highly dependent both on model assumptions and on the quality of input and monitoring data.

It is also crucial that the models are able to represent the processes involved at various temporal and spatial scales for the river basin under consideration. Moreover, the model should ultimately meet the user needs, demands and expectations. To this end, the selection of the appropriate model requires close interaction between policy makers, extension offices and basin managers and scientists.

In the case of this study, the objective was to assess the principal sources of nutrient loss into surface water and the impacts of agricultural mitigation measures and land use changes on water quality. The modeling tool used in the study included description of the physical processes and spatial variation of climate, soil, land cover, agricultural management and industrial and sewerage discharges. Often severe limitations on data availability required assumptions and data constructions. Consequently, this also hampered the possibility to comply with all the specific water manager and stakeholder interests.

In the Tungabhadra River, serious conflicts originate from the available water resources which are not enough to sustain all the needs that exist in the region for drinking water, agriculture (irrigation), fishery and industrial production. Not only is the quantity of the resource limited compared to the needs of the users, but also the quality is impaired due to non-sustainable use and less sound management. Several water works have been built in the last decades to increase water storage at catchment scale (dams, tanks) and to transfer water resources through canals to irrigated agriculture areas. The complexity of the landscape, climate and pollution sources – both point and diffuse sources – requires modeling tools to be used to help understand the complex interactions between the various sources and environmental processes.

With the modeling tools, as in this case, it is possible to estimate nutrient and sediment loads and to quantify the effects on water quality of downstream river reaches and/or reservoirs, including the identification of pollution sources along the river with the required source apportionment that is needed to rank critical areas.

These tools – as a single model or as chained sequence of different models – can also be used for: selecting suitable best management practices; ranking of sources and expected efficiency in nutrient reduction magnitude; and costs in addition to pollution impacts on various mitigation measures and scenarios.

The application of semi-distributed process oriented models for river basins (e.g., Arnold *et al.*, 1998; Krysanova *et al.*, 1998) have proven to be a good compromise between data availability and model complexity, in which the main processes are represented by physically based mathematical equations, while water/matter fluxes are expressed by simplified or conceptual formulations.

The Soil and Water Assessment Tool (SWAT) model was selected, being a continuous time simulation model aimed at river basin management. The choice of model was mainly based on the following facts:

- the model is comprehensive in simulating hydrology, crop growth, agricultural management, BMPs;
- it is a free and open source;
- it is widely used in the world and a large and a co-operative user community exists.

## 10.4 THE SWAT MODEL

SWAT is a basin scale, continuous time model that operates on a daily time step and is designed to predict the impact of management on water, sediment, and agricultural chemical yields in ungauged basins (Arnold *et al.,* 1998). It is currently being utilized worldwide in several large projects by government agencies and private consultancies as well as in research. The SWAT model operates on a daily time step and simulates eight major components: hydrology, weather, sedimentation, soil temperature, crop growth, nutrients, pesticides, and agricultural management. Major hydrologic processes that can be simulated by the model include evapotranspiration (ET), surface runoff, infiltration, percolation, shallow aquifer and deep aquifer flow, and channel routing (Arnold *et al.,* 1998). Simulation of the processes is accomplished via four subsystems: surface soil, intermediate zone, shallow and deep aquifers, and open channels. Streamflow in a main channel is determined by using three sources: surface runoff, lateral flow, and base flow from shallow aquifers. In SWAT, the impacts of spatial variations in topography, land use, soil, and other watershed characteristics on hydrology are considered in subdivisions. Watersheds are divided into sub-watersheds based upon drainage areas of the tributaries, and each sub watershed is divided into Hydrological Response Units (HRUs) based on land cover and soil type. Each HRU is assumed to be spatially uniform in land use, soil, topography, and climate. A daily water budget is established for each HRU based on precipitation, runoff, ET, percolation, base flow, and soil moisture change. Two options are available for partitioning surface runoff and infiltration: the SCS runoff curve number method (developed by the Soil Conservation Service of the USDA), and the Green and Ampt method (Green and Ampt, 1911). In the SCS method, a curve number is determined by land use, soil hydrologic group, and soil moisture conditions for each HRU within a watershed. The SWAT model offers three methods for estimating potential ET (PET): the Penman-Monteith model (Monteith, 1965); Priestley-Taylor model (Priestley and Taylor, 1972); and Hargreaves model (Hargreaves and Samani, 1985).

Sediment yield is calculated with the Modified Universal Soil Loss Equation (MUSLE) developed by Williams and Berndt (1977). The sediment routing model consists of two components, deposition and degradation, operating simultaneously. The deposition in the channel and floodplain from the sub-watershed to the watershed outlet is based on the sediment particle settling velocity. The settling velocity is determined using Stoke's Law and is calculated as a function of particle diameter squared. The depth of fall through a routing reach is the product of settling velocity and reach travel time. The delivery ratio is estimated for each particle size as a linear function of fall velocity, travel time, and flow depth. Degradation in the channel is based on Bagnold's stream power concept (Bagnold, 1977; Williams, 1980). Once the amount of deposition and degradation has been calculated, the final amount of sediment in the reach is determined.

The transformation and movement of nitrogen (N) and phosphorus (P) within a hydrological response unit (HRU) are simulated in SWAT as a function of nutrient cycles consisting of several inorganic and organic pools. Losses of both N and P from the soil system in SWAT occur by crop uptake and in surface runoff in both the solution phase and on eroded sediment. Simulated losses of N can also occur in percolation below the root zone, in lateral subsurface flow (including tile drains), and by volatilization to the atmosphere. Movement of nitrate ($NO_3$-N) in surface runoff, lateral subsurface flow, and percolation is computed as the product of the average soil layer $NO_3$-N concentration and the volume of water in each flow pathway. The mass of soluble P predicted to be lost via surface runoff is determined as a function of the solution P concentration in the top 10 millimeters of soil, the surface runoff volume, and a partitioning factor. Movement of organic N or organic and inorganic P on eroded sediment is estimated with a loading function initially derived by McElroy *et al.* (1976) and later modified for individual runoff events by Williams and Hann (1978). Daily losses are computed with the loading function as a function of the nutrient concentration in the topsoil layer, the sediment yield, and an enrichment ratio. In this study the model version SWAT2005 was used.

The ArcSWAT Interface has been created to facilitate pre-processing before running of the SWAT model and is known as ArcSWAT. The model requires a large amount of formatted inputs to be generated. The pre-processor is incorporated to handle all the inputs and also to graphically represent the model outputs after successful run of the model, as a post processing activity. The ArcSWAT interface consists of three segments, main interface, a preprocessor and a post-processor (Winchell *et al.*, 2007).

The main interface handles the creation of new SWAT projects, opening an existing project, copying an existing project, deleting an existing project. The pre-processor is the backbone of the interface. A SWAT model (run from executable file) requires extensive input files in their respective formats. The basic input required is the Digital Elevation Model (DEM) for the area under consideration. The pre-processor generates the 'Stream Network', identifies the outlet points for a given threshold value, delineates the main watershed and sub watersheds within it, if required. Watershed characteristics like area, slope, perimeter and channel characteristics are also calculated. Land use and soil grids are then overlaid and the basic modeling units are extracted. The other input files (soil, water use, management practices, pesticide, water quality etc.) for each sub-basin are then inputted. Default values are used in many files, which can be modified using the EDIT FILES menu. The sequence of input data creation is followed using the enable/disabled menu item. SWAT models are run by using the SWAT RUN menu.

The post – processor reads the results of the simulation run for the watershed as basin file and channel routing file in tabular form and helps in viewing the output created after the SWAT model has been run. The basin table and channel routing table are viewed (as daily, monthly, yearly frequencies).

## 10.5 MODEL IMPLEMENTATION AND DATA AVAILABILITY PROBLEMS IN THE TUNGABHADRA BASIN

To be able to run the SWAT model, it is necessary to utilize several geo-referenced databases which include the following: information on geomorphology (DTM, hydrology, geologic features); landuse; weather (temperature, rainfall, wind, radiation); management practices; point sources in the area; surface and groundwater flow; and water quality monitoring data.

Very often data availability constitutes a problem when using such complex and data demanding models; nevertheless some alternatives exist when detailed data are not available. In the case of the Tungabhadra, adequate soil type maps were missing, together with information on agricultural practices (actual crops, fertilizers and pesticide usage).

In the Tungabhadra modeling exercise, it was possible to resort to several 'global coverage' data sources, i.e.,:

- DTM from the NASA SRTM experiment;
- land use/ land cover from the Global Land Cover Facility (Hansen *et al.*, 2000);
- soil map based on FAO Global soils (FAO, 1995).

Examples of input data are given in Table 10.1 and Figures 10.1–10.4.

**Table. 10.1** Landuse in the Tungabhadra basin.

| Land Use | Area (ha) | % Watershed Area |
|---|---|---|
| Agricultural Land-Close-grown | 2550680 | 35.9 |
| Agricultural Land-Generic | 123459 | 1.7 |
| Agricultural Land-Row Crops | 29236606 | 41.1 |
| Forest-Deciduous | 476486 | 6.7 |
| Forest-Evergreen | 352598.29 | 4.9 |
| Forest-Mixed | 28187 | 0.4 |
| Summer Pasture | 374554 | 5.3 |
| Pasture | 192744 | 2.7 |
| Residential | 8104 | 0.1 |
| Water | 79123 | 1.1 |

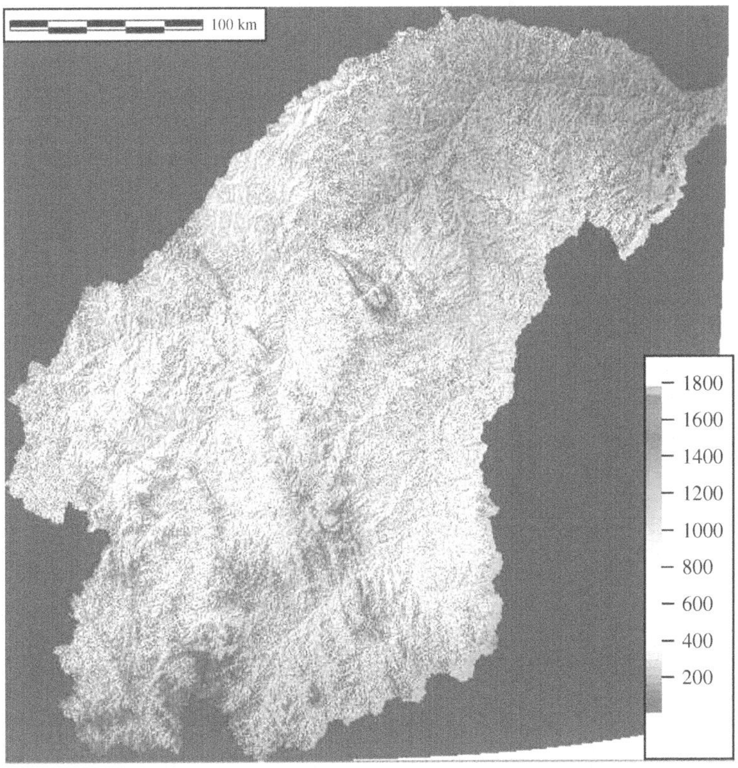


**Figure 10.1** SRTM Digital Elevation Model of The Tungabhadra basin.

Some data were only available at district level (i.e., actual crops, urban and rural settlements): in this case some statistical calculations were used to distribute this information within the delineated sub-basins.

Moreover, it was only possible to calibrate and validate the water flow. The water quality time series were incomplete, especially concerning the temporal resolution. Furthermore measurements on phosphorus and sediments were completely missing.

The Central Pollution Control Board (CPCB) and the State Pollution Control Boards (SPCBs) for the respective states of Karnataka and Andhra Pradesh are the only bodies responsible for the monitoring and control of pollution. Since it has proved somewhat difficult to get data from the Andhra Pradesh Pollution Control Board (APPCB), most of the data gathered deals mainly with the stretch of the river in Karnataka state, which in any case forms a major portion of it, and relies mainly on the data from the Karnataka State Pollution Control Board (KSPCB) and the Central Pollution Control Board (CPCB). However, there is considerable variation in the number of parameters monitored at the different stations over the years.

Quantitative daily data for 52 rain gauge stations and 8 temperature gauges are available for varying period from 1971–2004.

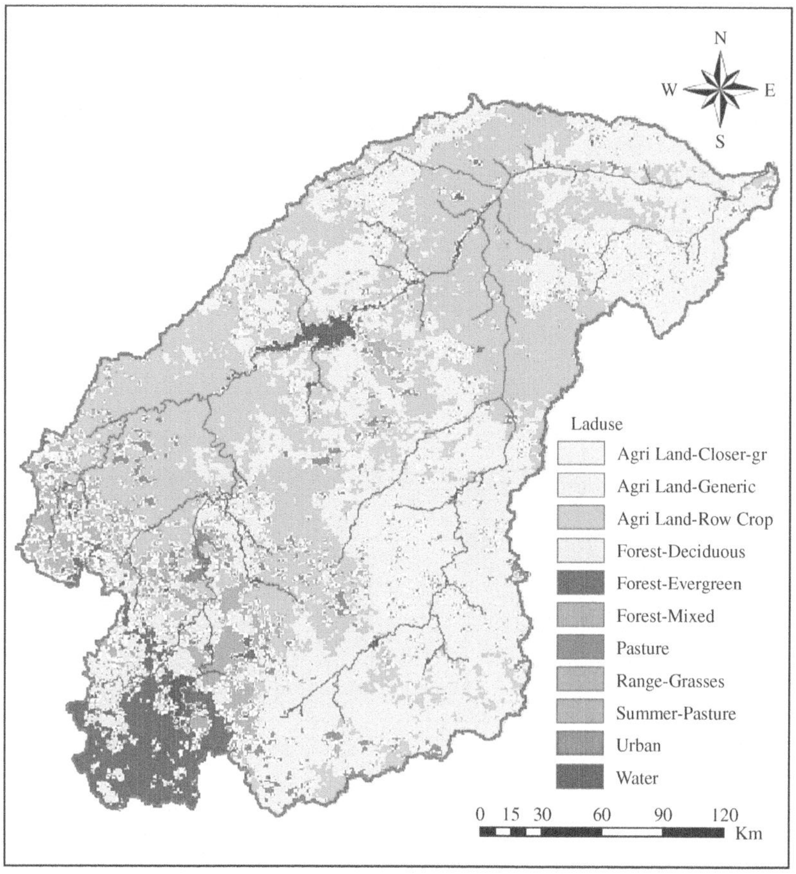


**Figure 10.2** Landuse of the Tungabhadra basin (Source: Global Landuse).

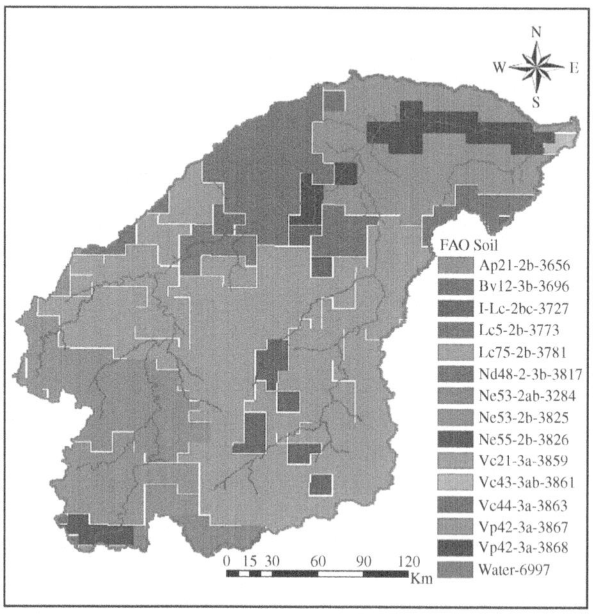


**Figure 10.3** Soils of the Tungabhadra basin (Source: FAO Global Soil).

The data generated in transient experiments by the 'Hadley Centre for Climate Prediction' U.K. at a resolution of 0.44° × 0.44° latitude by longitude grid points wasobtained from IITM, Pune (scenarios generated at Indian Institute of Tropical Meteorology, Pune using Met Office Hadley Centre regional climate model PRECIS). The daily weather data on maximum and minimum temperature, rainfall, solar radiation, wind speed and relative humidity at all the grid locations were processed. The Regional Climate Model (RCM) grid has been superimposed on the subbasins for deriving the weighted means of the inputs for each of the subbasins. The centroid of each sub basin is then taken as the location for the weather station to be used in the SWAT model. This procedure has been used for HadRM3 (PRECIS) baseline (1961–1990), A2 (2071–2100) and B2 (2071–2100) IPCC SRES scenarios with no sulfur.

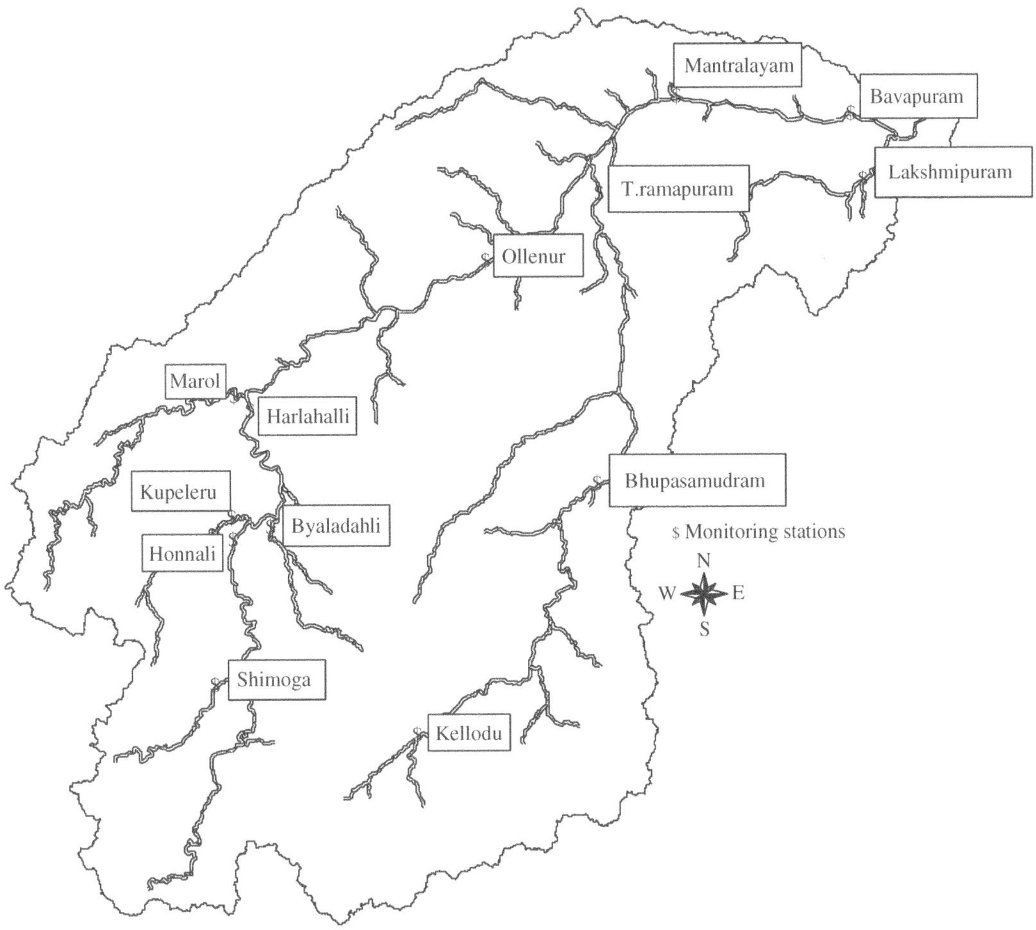


**Figure 10.4** Tungabhadra basin KSPCB water sampling stations.

The model has been run using the observed rainfall for the period 1971–2004 and also using HadRM3 SRES Baseline, A2 and B2 climate scenarios (30 years) without changing the land use. Though the model has generated very detailed outputs at the spatial and temporal scale, in the present analysis only some of the components that are considered important have been selected and reported.

The outputs of these scenarios have been analyzed with respect to the possible impacts on the water balance components such as runoff, soil moisture, actual evapotranspiration etc.

Although the model generates an exhaustively detailed output, the sub components of the water balance (those which have been identified to be used) for analysis are:

- total flow (water yield) consisting of surface runoff, lateral and base flow;
- soil water recharge (soil water recharge);
- actual evapotranspiration (actual ET).

These components are expressed in terms of average annual depth of water in mm over the total watershed area. In other words the total water yield is the equivalent depth in mm, of flow past the outlet of the watershed on average annual basis.

The annual average precipitation, water yield, actual evapotranspirarion and soil water recharge as simulated by the model over the Tungabhadra basin for the observed rainfall and the climate change scenarios are described in the following section.

During the modeling exercise some limitations and assumptions lead to the following uncertainties:

- detailed info on land use and soil type were not available;
- detailed data on the agricultural land use in the form of actual cropping pattern were missing;
- sewage contributions had to be calculated from literature data;
- point sources water use was assumed to be not consumptive;
- assumption made of no changes in land use in the modeling of future climate scenario.

## 10.6 RESULTS OF MODELING

The overall simulated water balance at the whole catchment scale is summarized in Figure 10.5 and Figure 10.6.

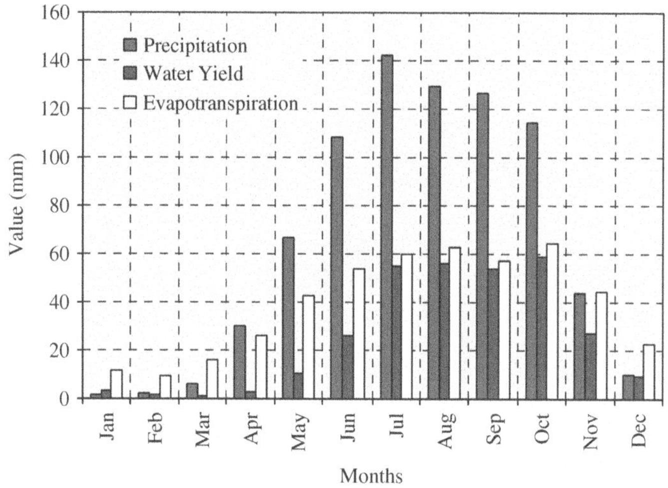


**Figure 10.5** Average monthly water balance components.

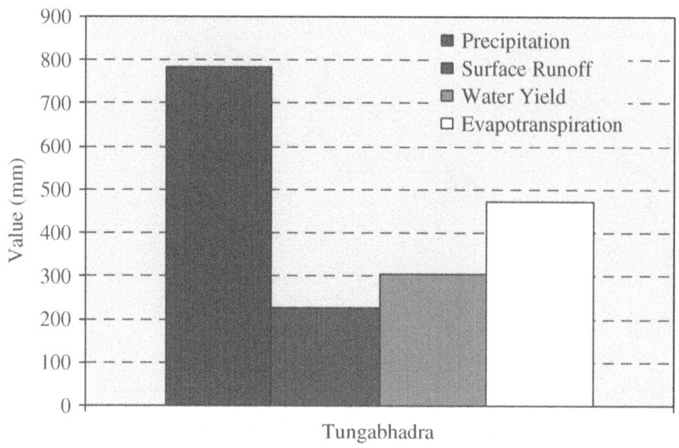


**Figure 10.6** Average annual water balance components.

The model has been calibrated (only hydrology) to obtain the best possible performance, notwithstanding the limitations mentioned above. Some examples of the results obtained are shown in Figure 10.7 demonstrating that the model captures the general behavior of the hydrographs with reasonable accuracy, which is also reflected in the high $R^2$-values. Nevertheless it is quite clear that the model tends to overestimate the water yield, probably due to the limitations encountered in describing the uses of water resources; possible withdrawals of water from the river have been omitted in the model inputs.

Maps summarizing some water balance elements at the sub-basin level are shown in Figures 10.11 to 10.14. Adopting a coarse discretization of the whole basin into 9 large sub-basins (for demonstration purposes) it was possible to show the gross accumulated water availability in the river basin (Figure 10.8), the average groundwater recharge (Figure 10.9 and Figure 10.12 at larger spatial resolution) and the average sediment yield (Figure 10.10).

It was also possible to document that the Tungabhadra dam drains the most sediment loaded sub-basin in the river basin (which are the highest in the south-west and north part of the basin; Figure 10.14), thus explaining the huge siltation rate recorded for the dam (22% volume lost due to siltation).

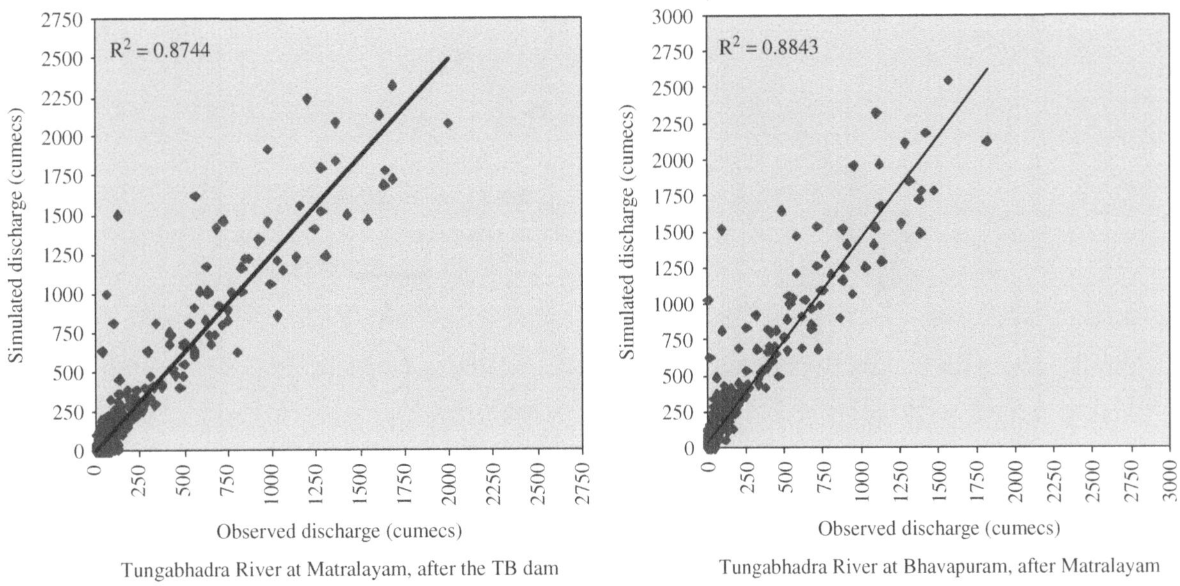


**Figure 10.7** Modeled vs. measured water flow data at two sites on the Tungabhadra River.

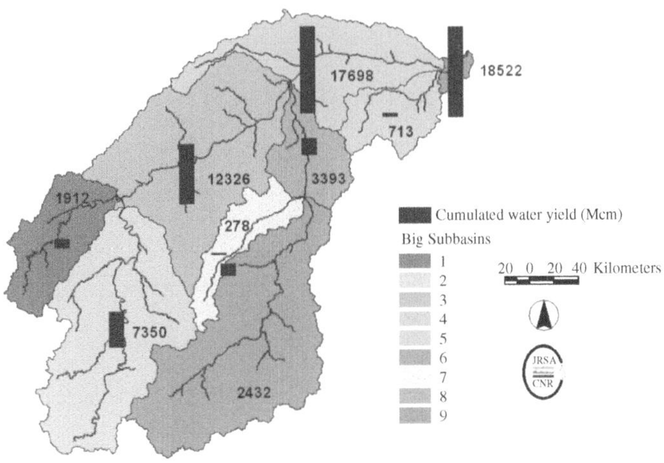


**Figure 10.8** Cumulated water yield.

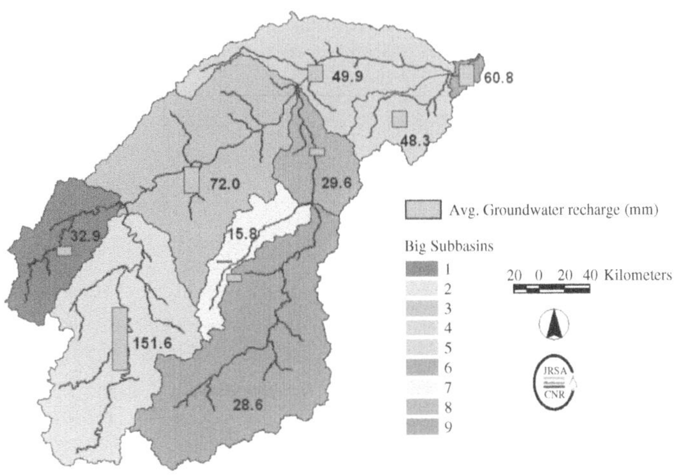


**Figure 10.9** Average groundwater recharge.

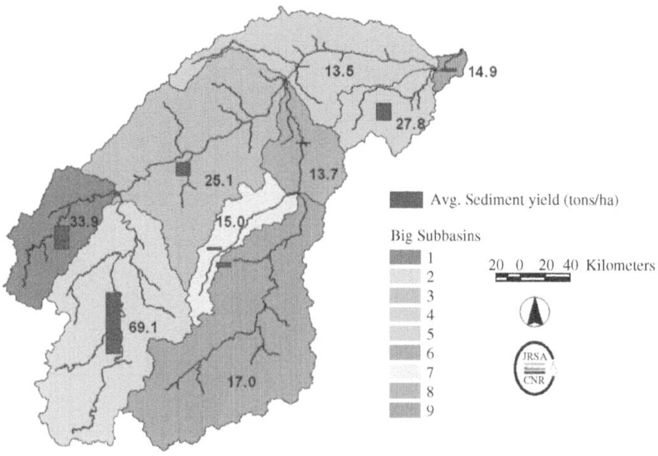


**Figure 10.10** Average sediment yield.

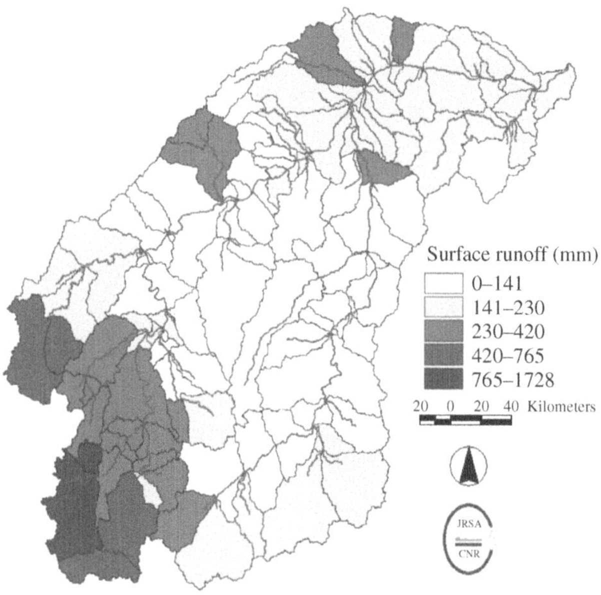


**Figure 10.11** Surface runoff.

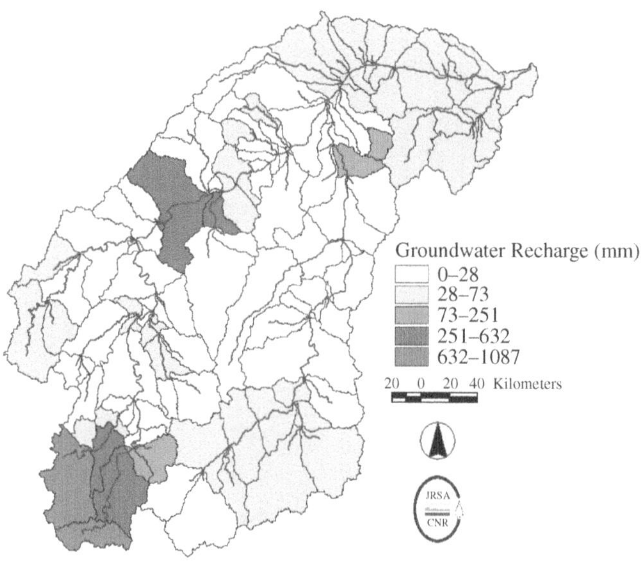


**Figure 10.12** Groundwater recharge.

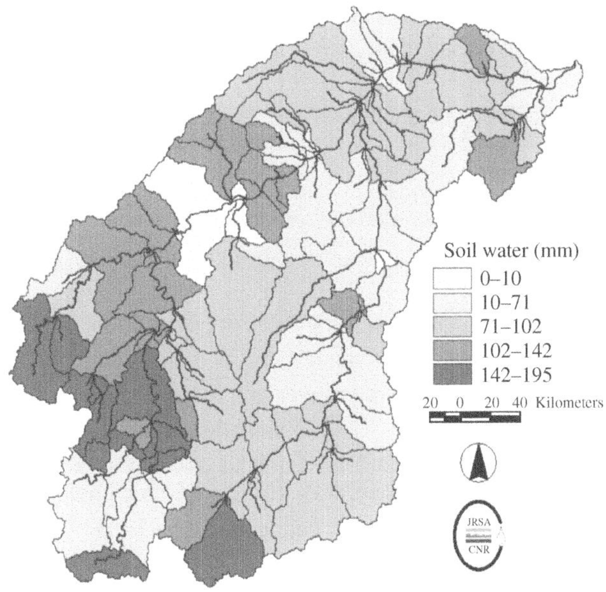


**Figure 10.13** Soil water availability.

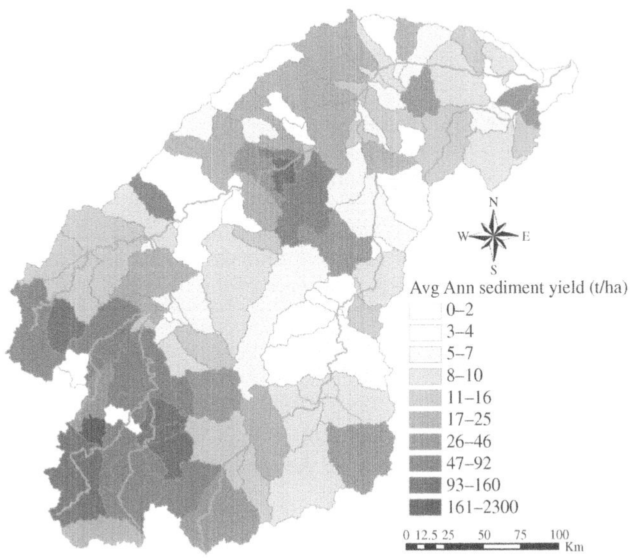


**Figure 10.14** Average annual sediment yield.

## 10.7 SCENARIO ANALYSIS

During stakeholder meetings, the participants were divided into break-out groups, each discussing the likely acceptable future changes. Scenario topics particularly addressed were: (i) water demand for different uses; (ii) land use and cropping pattern; (iii) agronomic practices; (iv) treatment of effluents and urban sewage; (v) sanitation practices in the rural areas; and (vi) trend in urbanization.

The discussed qualitative scenarios (in a 10-years perspective) were then translated into the following quantitative scenarios after agreement with the involved stakeholders (Table 10.2):

**Table 10.2** Quantitative scenarios agreed by stakeholders.

| Problem area | Projected change |
|---|---|
| **Industrial pollution** | strictly enforce existent laws and WWTP already existent; production accorded to river flow |
| **Mining** | surface runoff checked and avoided using infiltration ponds, check dams and management practices |
| **Sewage** | urban settlement 50% secondary treatment rural areas: 80% septic tanks |
| **Population growth** | 2.1%; urbanization trend: urban 60%, rural 40% |
| **Changes in crops** | shift from paddy to maize, sunflower and araca nut upstream of the TB dam |
| **SRI (System for Rice Intensification)*** | 40% shift in the tank area, 10% in the canal area |
| **Irrigation** | from submersion to sprinklers 80% change for araca nuts; no change for sugarcane |
| **Nutrient management** | 10% shift to organic farming for paddy and sugarcane |

(*) System of Rice Intensification is a method developed in recent years based on keeping soils moist but not saturated and transplanting young rice seedling in a wider way than traditional cropping method, allowing higher yields, less seeds requirement and less water for irrigation.

After discussions between stakeholders and modelers, the following scenario components have been considered for being simulated using the SWAT model (Table 10.3):

**Table 10.3** Shorter list of scenarios to be simulated.

| Problem area | Projected change |
|---|---|
| **Climate change** | Hadrm3 model, A2 emission scenario; 2070–2090 |
| **Sewage treatment** | 80% septic tanks in rural areas; 50% treatment in urban areas |
| **SRI** | 40% change upstream of the TB dam and 10% downstream |

After the model was calibrated and validated in the baseline simulation ($R^2 > 0.90$ NS $= 0.44$), it enabled us to apply the scenarios after having parametrized them (that is converting concepts into numbers used by the model).

However uncertainties present in the baseline simulation are reflected into scenarios output: the scenario on irrigation changes is not commented here because of uncertainties embedded in the cropping pattern and practices. Some of the results can be summarized, as in the following Table 10.4:

**Table 10.4** Summary results of the A2 climate change scenario.

| | Baseline | A2 | diff% |
|---|---|---|---|
| Precipitation mm | 829 | 928 | +12 |
| ET mm | 469 | 539 | +15 |
| Surface runoff mm | 177 | 219 | +24 |
| Water yield mm | 334 | 365 | +9 |
| Sediment load t/ha | 12.4 | 28.9 | +132 |
| Reservoir water inflow mm | 261 | 444 | +70 |
| Reservoir water outflow mm | 141 | 177 | +26 |
| Sediment inflow to reservoir t/ha | 0.6 | 1.9 | +253 |
| Sediment outflow from reservoir t/ha | 0.02 | 0.02 | 27 |

The climate scenario impact results indicate the following:

- a lower groundwater recharge rate is expected (Figure 10.15);
- a doubling in the sediment load (Figure 10.16);
- the sediment inflow into the TB dam is expected to increase three times while the sediment outflow will remain more or less stable; this will lead to a great increase of the siltation rate and a reduced life-time of the dam;

- rainfall increase of 12%;
- light increase in ET; small reduction in water yield;
- increase in surface runoff;
- diffuse source pollution slightly decreases.

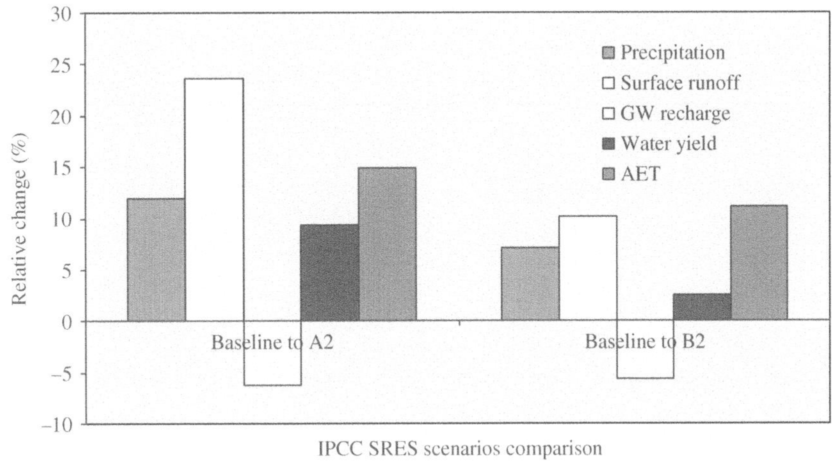


**Figure 10.15** Change in water balance components from baseline to A2 and B2.

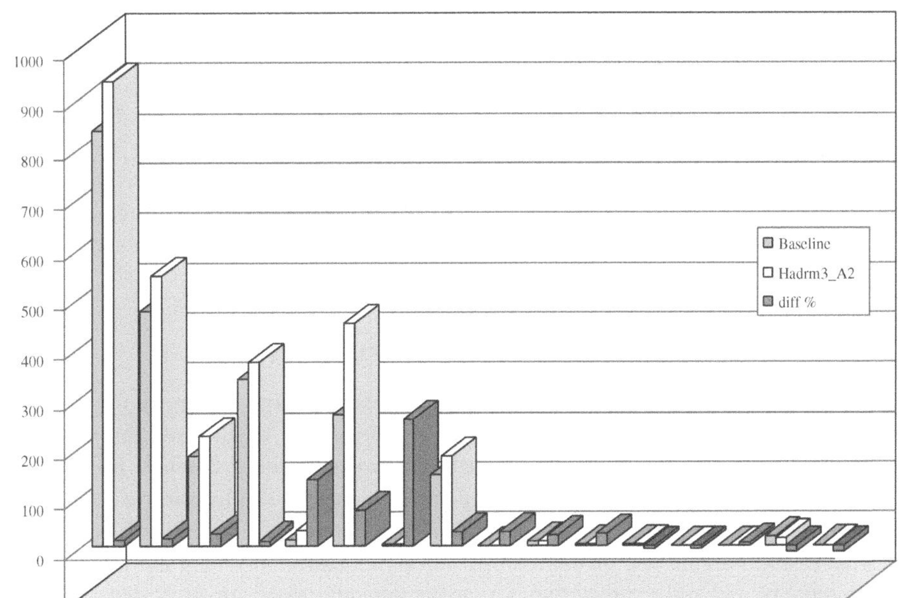


| | Precip mm | ET mm | Surf. runoff mm | Water Yld mm | Sedi-ment load | water inflow reserv | sed.inflow reserv | water out flow | sed. out-flow reserv | OrgN kg/ha | OrgP kg/ha | $NO_3$_sQ kg/ha | $NO_3$_rQ kg/ha | SolP kg/ha | $NO_3$_gw kg/ha | P_gw kg/ha |
|---|---|---|---|---|---|---|---|---|---|---|---|---|---|---|---|---|
| Baseline | 829 | 469 | 176.56 | 334 | 12.4 | 261 | 0.6 | 141 | 0.0 | 7.3 | 0.8 | 1.2 | 0.1 | 0.0 | 16.4 | 0.3 |
| Hadrm3_A2 | 928 | 539 | 218.72 | 365 | 28.9 | 444 | 1.9 | 177 | 0.0 | 8.8 | 1.0 | 1.1 | 0.1 | 0.0 | 14.2 | 0.2 |
| diff % | 12 | 15 | 24 | 9 | 132 | 70 | 253 | 26 | 27 | 21 | 24 | −8 | −6 | 5 | −14 | −15 |

**Figure 10.16** Anticipated effect of climate change on selected parameters (entire basin; scenario A2).

Some conclusions can be drawn regarding the sewage treatment scenario. The sewage contribution to the overall river pollution is moderate as for nutrients, huge for pathogens (Table 10.5).

Prudent estimates on treatment efficiency and implementation lead to big reduction of nutrient load summarized in Table 10.6. The simulated sewage treatment scenario showed that a 50% reduction from sewage discharge can be expected in both N and P contribution to the overall river pollution.

**Table 10.5** Ratio (in percentage) of the sewage vs. diffuse sources contribution.

| | |
|---|---|
| Tot N | 5.2 |
| Tot P | 26.2 |
| Sediment | 15.4 |

**Table 10.6** Summary of the results from the sewage treatment scenario simulation.

| | | | |
|---|---|---|---|
| Tot N (ton/day) | 7,1 | Tot N (ton/day) | 3,5 |
| Tot P (ton/day) | 2,3 | Tot P (ton/day) | 1,1 |
| Sediment (ton/day) | 203,3 | Sediment (ton/day) | 135,7 |
| Untreated load | | Treated load | |

The simulated implementation of System of Rice Intensification (SRI) was estimated to lead to an expected 6% saving of the river basin water yield, accompanied by an overall increase of around 20% of rice yield. However, the expected (imposed) change in water usage in paddy cropping (on 10% of the surface of the whole basin) seems to be too small compared to the actual average water use in agriculture.

As a general comment on the quality of the scenario simulation exercise, given the described constrains in data availability, it can be stated that even if the model was calibrated only for the hydrology, scenarios simulations are considered reliable since:

- climate change impact simulations have been only run for hydrology (and for erosion that is greatly driven by runoff);
- calculated actual erosion rate is in agreement with measured siltation rate in the Tungabhadra dam (since monitored sediment transport is missing);
- water saving due to SRI diffusion scenario is based on changes in irrigation and then mostly on hydrology;
- sewage treatment scenario is based on simple point source load calculations.

However uncertainties in modeling results arise due to uncertainties regarding water withdrawals from the streams and groundwater, as these are estimates based on official data and expected standard water consumption for crops in the area. Also uncertainty exists concerning the application of agrochemical applications as these also are based on official statistics and census data.

## 10.8 CONCLUSIONS AND MANAGEMENT IMPLICATIONS

The use of a basin scale, continuous time model can help in gaining a better understanding of the water, sediment and nutrient balance in a river basin, allowing critical source areas to be singled out where mitigation measures (management policies) need to be carried out. In such a process, scenarios which have been developed through interaction with stakeholders, can be simulated to help in the decision making process, allowing the assessment of suitable best management practices and providing the ability to rank them on the basis of expected efficiency.

To ensure profitable and reliable use of modeling results, the availability of good quality data regarding climate, water quantity and -quality and soil characteristics is a fundamental requisite. It is necessary to put more efforts in planning effective monitoring activities: more frequent water quality sampling, a larger selection of parameters to be measured, namely phosphorus, ammonia and suspended solids. The actual high level of erosion and increase in simulated trend in the future due to climate change suggest that erosion control should gain a high priority in next future landscape management policy.

Investment in improved sewage treatment in urban and rural areas will significantly reduce the nutrient load to the river and can dramatically decrease the level of pathogens in surface water.

Active policies must be undertaken with a deeper effort to help the diffusion of new techniques in rice cropping (SRI) that are likely to play a relevant role in saving water resources.

## 10.9 REFERENCES

Arnold, J.G., Srinivasan, R., Muttiah, R.S. and Williams, J.R. (1998) Large area hydrologic modelling and assessment part I: model development. *Journal of the American Water Resources Association* **34**, 73–89.

Bagnold, R.A. (1977) Bed-load transport by natural rivers. *Water Resources Research* **13**, 303–312.

DMG (2004) Report on dynamic ground water resources of Karnataka as on March-2004, Department of Mines and Geology, Government of Karnataka and central ground water board South Western region, India.

Food and Agriculture Organization of the United Nations (FAO) (1995) Digital Soil Map of the World (DSMW) and Derived Soil Properties. Version 3.5. CD-ROM.

Green, W.H. and Ampt, G.A. (1911) Studies on soil physics, part I, the flow of air and water through soils. *J. Agric. Sci.* **4**(1), 1–24.

Hansen, M.C., Defries, R.S., Townshend, J.R.G. and Sohlberg, R. (2000) Global land cover classification at 1 km spatial resolution using a classification tree approach. *Int. J. Remote Sensing* **21**(6&7), 1331–1364.

Hargreaves, G.H. and Samani, Z.A. (1985) Reference crop evapotranspiration from temperature. *Applied Engrg. in Agric.* **1**, 96–99.

Krysanova, V., Muller-Wohlfeil, D.-I. and Becker, A. (1998) Development and test of a spatially distributed hydrological/water quality model for mesoscale watersheds. *Ecological Modelling* **106**, 263–289.

McElroy, M.B., Elkins, J.W., Wofsy, S.C. and Yung, Y.L. (1976) Sources and sinks for atmospheric N2O. *Rev Geophys Space Phys.* **14**, 143–150

Monteith, J.L. (1965) Evaporation and the environment. In *The state and movement of water in living organisms. 19th Symposia of the Society for Experimental Biology*. Cambridge Univ. Press, London, UK p. 205–234.

Priestley, C.H.B. and Taylor, R.J. (1972) On the assessment of surface heat flux and evaporation using large-scale parameters. *Mon. Weather Rev.* **100**, 81–92.

Williams, J.R. and Berndt, H.D. (1977) Sediment yield prediction based on watershed hydrology. *Trans. ASAE*, 1100–1104.

Williams, J.R. and Hann, R.W. (1978) Optimal operation of large agricultural watersheds with water quality constraints. *Texas Water Resources Institute, Texas A&M Univ., Tech. Rept.* No. 96.

Williams, J. R. (1980) SPNM, a Model for Predicting Sediment, Phosphorus, and Nitrogen from Agricultural Basins. *Water Resources Bull.* **16**(5), 843–848.

Winchell, M., Srinivasan, R., Di Luzio, M. and Arnold, J. (2007) ArcSWAT interface for SWAT2005 – User's Guide. Blackland Research Center, Texas Agricultural Experiment Station and Grassland, Soil and Water Research Laboratory, USDA Agricultural Research Service, Temple, Texas.

# Chapter 11

# Tungabhadra sub-basin: Recommendations

*K.J. Joy, S. Paranjape, S. Manasi, P. Mollinga and U.S. Nagothu*

## 11.1 THE BASIN CONTEXT AND ISSUES

The preceding three chapters in this section of the book, which focused on the Tungabhadra sub-basin (TBSB) discuss the main critical issues in the basin from an IWRM perspective. The key message from these chapters is that IWRM has been mostly an issue of rhetoric and that it has not made much headway in terms of the real management of the basin.

The TBSB includes a wide variety of state actors and non-actors or interest groups which have very little interaction between or amongst them. The state actors are the various government departments or autonomous corporations that deal with TBSB water management at different levels – national, state, district and local levels. The most important departments are the Water Resources Department of Karnataka and Andhra Pradesh states, earlier named the Irrigation Departments, which are technically responsible for water resources and their allocation across different uses and users including storage and conveyance of water, pricing and maintenance. The State Pollution Control Boards (SPCB) of Karnataka and Andhra Pradesh states with their branches at the district level are important state actors as they are ultimately responsible for water quality and monitoring and control of pollution. Besides the different state agencies looking after sewerage and supply, drinking water, smaller water bodies (smaller water bodies are looked after by the self-government bodies like municipalities, zilla parishads and panchyats), groundwater, water infrastructure and construction, etc. there are many others. There also appears to be a reluctance to share data across the two riparian states and even across different departments within a state, with each department functioning in isolation often leading to turf wars between different departments and departmental fundamentalism! It is only in recent years that initiatives are being taken to bring the departments together, and make data accessible, as observed in Andhra Pradesh.

The non-state actors or stakeholders also comprise of the direct users of water (farmers, fishermen, rural artisans who directly depend on the water resources for their livelihoods, as well as the rural and urban population who use water for drinking, domestic use and for cattle and industries) along with organizations and interest groups like Water Users' Associations (WUAs), academic and research institutions engaged in the study of water and water use, and different types of civil society organizations that engage with water issues and water rights. The direct users have limited access to data; and not involved in generating knowledge and the formulation of policies. There appears to be no legal and institutional space for stakeholder participation at the moment. While consultative meetings are sometimes held with NGOs and representatives of various interest groups and organizations there seems to be no guarantee that their viewpoints will be taken into account while formulating the policies.

In India, there are in many cases a polarized relationship between the state and civil society organizations (CSOs) as well as between the CSOs and industry with mistrust and hostility across the divide. In the TBSB in particular, the relationship between the industry and CSOs has been one of polarization because of misgivings and the deep-seated public perception that industry is the main contributor to the Tungabhadra river pollution. There was a concerted civil society action in Karnataka state on the issue of industrial pollution and at the CSOs' initiative a Tungabhadra Watchdog

 *Integrating Water Resources Management: Interdisciplinary Methodologies and Strategies in Practice.* Edited by Geoffrey D. Gooch, Alistair Rieu-Clarke and Per Stålnacke. ISBN: 9781843393252. Published by IWA Publishing, London, UK.

Committee (TWC) has been formed and it works very closely with the SPCB and has also succeeded in getting access to the monitoring data.

Although civil society actions seem to have had a positive impact on the pollution levels in the river, the downside is that representatives of industry now tend to look at any effort, even academic exercises like the STRIVER project, suspiciously and this was likely the main reason why industry did not participate in the stakeholder meetings organized as part of the project. Policy formulation is primarily a political process and science does not play a key role in this. The STRIVER project tried to address two important issues in TBSB through more focused research: 1) land-water use interaction including the question of inter-sectoral water allocation (or competing water uses); and 2) water quality and pollution. From an IWRM perspective both these issues are very critical and are also relevant to TBSB. In fact, during the first stakeholder meeting the stakeholders also listed these two as the most critical issues in the basin. However, stakeholders also highlighted other important issues in the basin and they include: lack of inter-sectoral and inter-departmental integration or coordination; absence of a river basin organization (and as a consequence there is lack of an integrated basin management plan based on IWRM principles even though the national and the state level water policies emphasize river basin agencies and plans); lack of legal and institutional space for stakeholder participation in governance issues; non-availability and access to reliable data at appropriate scales to the public; and absence of a well thought out strategy for training and capacity building of stakeholders (Society for Promoting Participative Ecosystem Management and International Water Law Research Institute, 2007).

In this chapter our intention is to provide some important recommendations that are seen as necessary if TBSB is to move towards IWRM in the near future; these recommendations are based on the studies carried out in the STRIVER project. Of course, here IWRM is seen more as a 'boundary concept' (Mollinga, 2006) with enough flexibility to take into account the specific bio-physical, economic, socio-cultural, political-institutional-legal context of a particular basin or nation state. The recommendations are organized under the following broad groupings: (1) inter-sectoral integration; (2) source integration; (3) departmental and institutional integration including stakeholder participation; (4) access to data and information; and (5) training and capacity building. We have made substantial use of the different policy briefs and work package deliverables in this chapter, generated as part of the STRIVER project. They are available at, http://www.striver.no or http://kvina.niva.no/striver/STRIVERBriefs/tabid/100/Default.aspx.

## 11.2 RECOMMENDATIONS

### 11.2.1 Inter-sectoral integration

In IWRM one of the important issues that needs to be addressed is the multi-functionality of water in river basin planning. Water use in a river basin is often determined by the livelihood patterns in the basin. Also, multi-functionality includes the needs for different livelihoods (farming, fisheries), environmental flows, water for irrigation, drinking water, cultural and spiritual uses for instance. Currently, there is a very little arrangement in place to integrate the water needs of different sectors at the basin level. The allocation and management of water is not only determined by water demand and supply, but also by the political obligations. One of the basic recommendations in this chapter is to develop control over water resources in terms of the quantity available in order to ensure the minimum needs of various sectors. Hydrological modeling of the sub-basin could be useful in this context. The planning documents related to TBSB (in fact these documents are more project specific, rather than viewing the river basin as a whole and that in itself is a weakness) do not include a description of the impact of actions, for example, of changes in river water flow on species, ecosystems or livelihoods patterns. In the TBSB, a water scarce basin, conflicts are related to water availability (or allocation of water) for different water use sectors leading to competing interests between agriculture and urban water supplies or industrial uses leading to different types of water conflicts. In the TBSB there are also conflicts due to high demand for water between upstream and downstream farmers as more water demanding cash crops are planted (Manasi *et al.*, 2008; Manasi *et al.*, 2009).

Another related issue is the management of water in TBSB. It is managed at the state level, and not at the basin level. Within the state different water abstractions and uses are managed by sectoral departments. This has led to a fragmented planning and management of the water resources and the existing allocation system fails to provide for variation resource availability, or take into account the uncertainties caused by climatic changes.

It is important to adopt IWRM for improved water management in order to prevent further degradation and to promote sustainable water use. It is expected that IWRM will become more relevant as climate variation and extreme events such as droughts become more frequent in basins such as TBSB. If this is to happen, water management within the agricultural sector needs to address a wide range of issues, management of supplies and demands, increasing efficiency of water use, balancing competing demands, sustainability of agro-ecosystems and other water-dependent ecosystems. Some of the specific recommendations for inter-sectoral integration from the Striver research team are:

- Existing policies should be revised to address the current and future water scenarios in the basin, especially the regulatory mechanisms for implementing and enforcing them, are limited or non-existent.
- There is a need to take initiatives to develop an overall basin wide IWRM plan addressing multi sectoral and interstate needs, balancing competing water demands and taking the supply and demand, climate change impacts and water pricing into account.
- Promote technologies and innovations with greatest impact on water use efficiency and livelihood assurance.
- Water use efficiency should be improved through promotion of options like System of Rice Intensification (SRI) in irrigated agriculture through government programs and promoting fisheries in the basin. Various drip irrigation systems are available for growing sugarcane and orchard crops. However, these methods are only partially adopted (SOPPECOM, 2008).
- The efficiency of industrial water use must be improved through proper pricing and providing credits for reuse and recycling and strict enforcement of effluent quality standards.
- Planned integration of policies and programs should be undertaken for rain fed and irrigated agriculture (Mollinga *et al.*, 2008).

In the TBSB inter-sectoral integration is also important from the point of view of different livelihoods. The study showed that nearly 10000 households in the basin depend on fisheries for their livelihoods. These are basically poor and small scale fishermen. Despite this dependency, the Water Resources Department does not consider fisheries as one of their priorities when allocating water. The State Water Policy does mention that water should be ensured for various sectors and livelihoods, but does not specify how this policy should be implemented. The development initiatives of the Fisheries Department are not integrated with the activities of the water resources department or the Agriculture Department. Overall, there appears to be a declining trend in fisheries production accompanied by a change in the institutional mechanisms that is not in the interests of the small scale fishermen. Integrating water management with the needs of other sectors such as fisheries, will not only ensure livelihoods of marginal communities such as the small-scale fishers, but also increase the water use efficiency in the basin

A number of institutional measures could be implemented to improve fisheries in the basin. For example, there is at present a lack of integration between the relevant polices, departments (state and local agencies) and programs that need to be pulled together to facilitate the access of the poor to fisheries. The government should reconsider issuing fishing rights to small scale and traditional fishermen rather than open tenders, customize training programs and implement government welfare schemes to benefit the poorest amongst the small scale fishers. For example, the comprehensive policy for auctioning fishing rights in tanks, reservoirs, and rivers was formulated in 1997 and later amended in 2005, by the Karnataka Government (Nagothu and Manasi, 2008a,b). As per these amendments, the tanks were supposed to be leased out to the WUAs formed under Jalasamvardhana Yojana Sangha, or the registered Fisheries Cooperative Societies or the registered water users' cooperatives and other societies/institutions, in order of priority. However, policy needs to be implemented properly to ensure that the local fishermen or their societies benefit. A number of measures can be initiated at the local level, for example, improving the fisheries in the water bodies, issuing licenses only to small scale and traditional fishermen, customizing training programs to suit various types of fishermen and implementing government welfare schemes to benefit the poorest amongst the small scale fishers. Fishermen are concerned about the fishing rights in the various water bodies. This is up to the government agencies to decide how the rights will be granted in the future, and whether it is the welfare of the poor that is important, or earning revenue from fisheries (Nagothu and Manasi, 2008a,b).

Another important area that needs to be considered is the issue of water quality and pollution. The STRIVER study clearly shows that water quality is a serious issue in TBSB (see Chapter 10) and the sources of pollution could be apportioned to industrial pollution, untreated sewage from the municipalities of the towns and cities and non- point pollution from agricultural chemicals. One of the important issues that came up during the study, especially for modeling, was the issue of data availability. Thus one of the important issues that need to be addressed in this context is to have proper data collection – at appropriate scales spatially and temporally – and its availability.

Some of the recommendations, especially from a management perspective, could include (Lo Porto *et al.*, 2009):

- more efforts in planning effective monitoring activities, particularly water quality sampling strategies, data collection, analysis and recommendations based on results;
- make erosion control a primary goal for landscape management;
- improving the efficiency of industrial use through proper pricing and providing credits for reuse and recycling and strict enforcement of effluent quality standards are required.

### 11.2.2 Source integration

In India one of the main areas of polarized discourse is around water source creation; if we are to ground IWRM in India then this polarization especially between large and small-scale users and storage systems, needs to be bridged. The mainstream strategy so far has given preference to large water storage systems, i.e., large dams, rather than smaller and dispersed storage systems or micro-watershed based approaches. Conversely, most civil society organizations (CSOs), especially those working on issues of environment and displacement have been opposing, time and again, the building of large dams. "Partly as a consequence of the extreme and inflexible positions around source creation, some of the fundamental issues in the water sector have been left unaddressed. Most importantly, it is not recognized that there now exist technically sound, alternative methods of designing or restructuring large water resource development projects in a way that ensures sustainability and equity, makes them compatible with the development of local sources and participative management, and greatly reduces the human and environmental costs of the projects" (Paranjape and Joy, 1995, p.8).

Similarities are found regarding surface and groundwater. Though the 'conjoint' use of surface and ground water is mentioned in policy and project documents, yet it has not yet been made part of the design of projects and they operate as independent systems. Government program seem to follow this fragmented approach to water resource development and water use. The government programs related to micro watershed development and groundwater recharge, renovation of the tanks and the efforts at larger reservoirs is not integrated and the issues are planned and implemented separately under separate departments.

The study of tanks carried out in the upper part of the Tungabhadra catchment (Chapter 9) showed that tanks are complex, multifunctional entities that support villagers' livelihoods in a number of ways. Their rehabilitation needs to be informed by an IWRM perspective that takes care of their multi-functionality (as tanks have various direct and indirect uses as already discussed in Chapter 9) as well as relationships with larger sources. Tanks, therefore, have great potential of taking us towards IWRM and widely dispersed livelihood assurance provided we have a suitable policy aimed at realizing this potential.

The study described in Chapter 9 also clearly highlights the need for tanks to receive stabilizing supplements from larger, more assured sources to gain better performance particularly in respect of local rainfall variability. Many studies show that rain-fed tanks show good performance for years with better rainfall, but may not perform very well in bad years. Also, studies (Vaidyanathan, 2001) show that the performance of system tanks in Tamil Nadu (tanks which receive regulated supplements from larger systems) is better than other tanks. STRIVER's study of tanks in the Tungabhadra basin also shows that balance reservoirs (tanks that serve as a buffer storage for extending canal reach) perform much better than the rain-fed tanks in terms of taking care of shortages (Paranjape *et al.*, 2009).

The following suggestions accommodate this integration of different types and scales of sources or waterscapes in the TBSB (Paranjape *et al.*, 2009):

- all tanks with a net irrigated area up to 200 ha, the traditional size of tanks) should be governed by tank-related IWRM institutions;
- service area and/or a users' list should be worked out for each tank on the basis of its multi-functionality;
- adequate representation should be provided to non-irrigation users of tanks, especially those who may not own land in the designated commands (this would ensure a representation from all groups within the villages served by the tank);
- adequate representation should be provided for Panchayat Raj Institutions (PRIs) as PRIs largely do and should deal with other water use related programs;
- the governance and management of tanks should be integrated with other water uses (drinking water, sanitation, fairs, water for cattle);
- rehabilitation should go beyond desiltation and should ideally precede the transfer to participative institutions. This aspect is crucial in the social sustainability of tank rehabilitation, especially in respect of groups that were excluded from tank benefits in traditional systems.

### 11.2.3 Institutional and policy integration and stakeholder participation

Institutional framework is a prerequisite for both inter-sectoral integration as well as integration of sources across different types and scales. It is also important in the transition towards IWRM. The National Water Mission released by the Government of India in 2008 does mention the possibility of a river management board and integrating management efforts at the basin level (www.indiaenvironmentportal.org.in/). However, this demands sufficient institutional support at various levels. However, as the STRIVER study shows, this has been one of the major weaknesses in TBSB. Firstly, there are too many departments dealing with different aspects of water with very little coordination between them. Secondly, although participation is mandated in many of the water related policy instruments and constitutional efforts have been made at decentralization (efforts like the 73rd and 74th constitutional amendments), there is no policy or legal mandate for stakeholder consultation and involvement. Thirdly, there is the absence of a comprehensive policy framework that engages with all critical aspects of water. Something akin to the European Water Framework Directive does not exist in India. Fourthly, there is the polarized relationship between the state and CSOs, as well as between the CSOs and industry (as already discussed above). In the case of the TBSB, as in the case of almost all trans-boundary or inter-state rivers in the country, there is no basin level organization; nor is there an organization for the portions of the basin that fall within the two riparian states of Karnataka and Andhra Pradesh. Thus, the main policy thrust should be to bring in a comprehensive water policy and/or law that covers all areas and addresses issues such as integrated planning and management at basin level and integration of sources of different types and scales. Furthermore, institutional integration should take the form of a nested (federated) institutional framework starting from the micro-watersheds, moving upwards and finally covering the whole basin. This should also include the integration of sectors, norms for allocations for different sectors including the ecosystem needs (environmental flows), water quality issues, stakeholder participation, access to data and information and training and capacity building. Some of the specific recommendations in Manasi *et al.* (2008) and Manasi *et al.* (2009) are:

It is recognized that it is more important to move towards greater integration between current uses and institutions in a nested institutional framework. However, it is important that the nested institutions do not stop at the state boundaries, but also cover the entire river basin in the form of a river basin organization with adequate representation from all important stakeholders. A beginning could be made by forming two organizations for the basins situated in the two states of Karnataka and Andhra Pradesh. This would, at the moment, be more feasible politically. Then an interstate forum (or river basin organization) of state level basin organizations could be established providing continuous interaction at the basin level with regard to, for instance, interstate allocation, scheduling flows, monitoring water quality and environmental flows and anticipating and effectively dealing with impending disasters such as floods and droughts.

It is also considered necessary to initiate a participatory process of reforming the water laws and water institutions through extensive consultations involving affected agencies and public in order to balance requirement and availability across space and time and to provide adequate institutional and policy support to clarify the entitlement and responsibilities, roles, allocations, legal status of all users and water providers. This could ensure sustainable use and integration of responsibilities of water management between departments with a formal mechanism to ensure co-ordination.

It is important to establish a regulatory body and regulatory mechanism that provides for dispersed access to the authority and/or its officials. Particular attention should be given to supporting women and marginal communities and providing them adequate access to the regulatory authority in defense of their rights and entitlements as well as for redressing disputes over water entitlements, allocations and tariffs. It is also necessary to provide for mandatory periodic presentation of state level water resource plans for all sectors in the basin, including allocations and tariffs, after due consultation and representation from all stakeholders.

Furthermore, it is necessary to provide legal and institutional space for stakeholder participation and also make sufficient budgetary allocations to support the activities of incorporating stakeholder interaction into IWRM at the basin level. The move towards greater regulation by independent bodies must be matched by a clearer definition of entitlements for the different sections and uses. A clearer set of guidelines are also required for balancing entitlements from a rights perspective and those from an economic goods perspective.

### 11.2.4 Access to data and information

Access to reliable data and information is important for integrated planning of the basin and monitoring the changes and for supporting informed participation by the different stakeholders. Since stakeholders as a group are heterogeneous and unequal in terms of power and access to information, those stakeholders who are closer to the state such as the different state departments, or those who are more powerful and resourceful like the industries, have easier access to information. Other stakeholders, such as farmers, very often do not have ready access to data and information. In fact, if these stakeholders are to participate on an equal footing with the more powerful ones then there is a need to create a level playing field for information. The problems are of twofold, reliability and accessibility. There is an overall ambiance of secrecy with regard to data, especially data on upper portions of the basins which consist of areas which are frequently left under-protected. Though most of the data is collected by publicly funded institutions, very often the data is either not made available within the public domain or the public have to pay exorbitant fees to purchase it. Each of the states collects and maintains its own data and hence it is rather difficult to get basin level data from one source. Evidently, the riparian states maintain different sets of data and during negotiations between the states each of them bring forth or use the data sets that are more advantageous to that particular state. The Right to Information Act in India was meant to provide access to information but from the experiences over the last few years of its implementation one can say that it has only been partially successful. Some of specific suggestions to improve the situation include:

- the development of an appropriate basin level land and water use database for developing a holistic perspective and carrying out a complete hydrological characterization to prepare an account of the water resources available and of their regularity in time and space;
- that data should be collected and maintained by independent agencies;
- that data should be made available to all the stakeholders in an easily accessible and understandable manner and put all the data on the public domain.

### 11.2.5 Training and capacity building

Capacity building is needed to prepare water users, water managers and policy makers for adopting an integrated approach to water management. Though this has been a neglected aspect in the basin, recently there is increased awareness of the need for capacity building and though efforts are fragmented, there is some movement in this direction. For example, in the area of participatory irrigation management there are provisions for the training of the farmers, office bearers of the Water Users' Associations (WUAs) and also of the departmental officers, by visits to successful case areas. Similarly, in the area of micro-watershed development and self-help groups (micro credits) very positive experiences are evident. However, the funds allocated by the government are very meager and there is the need for an adequate budget for the training of water users and managers at different levels.

In both the riparian states of Karnataka and Andhra Pradesh there are Water and Land Management Institutes (WALMIs) which are supposed to be nodal agencies for taking up training and capacity building. These institutions should be supported regularly with updated information and linked to the scientific agencies for developing courses for different categories of water users and managers. There are also other academic institutions like the universities (especially agricultural universities) who are involved in training. The agricultural departments often have a fairly elaborate 'extension' system providing various inputs, including knowledge inputs to farmers.

It is rather unfortunate that none of the mainstream academic institutions offer a full-fledged academic course on IWRM, although the South Asia Consortium for Interdisciplinary Water Resources Studies (SaciWATERs) over the last few years have made an interesting effort to introduce an IWRM course at the masters level in a few engineering colleges/universities in South Asia (for details visit www.saciwaters.org). There are however also many NGOs who are involved in training and capacity building in the area of natural resource management, governance issues, livelihoods, and micro-credit loans.

If IWRM is to be grounded in TBSB there is an urgent need to learn from these efforts, even though they are fragmented, and come up with a clearer strategy for training and capacity building at various levels and for different stakeholders. Here again the various agencies and institutions need to come together and pool their experiences and expertise to formulate a comprehensive program for training and capacity building. There could be a nodal agency at the TBSB level, or two nodal agencies for the two different states and perhaps in this respect the existing WALMIs could take the lead in this? Efforts also need to be made to facilitate the participation of NGOs into this type of program, as if the state agencies do not take into account the NGO experience and there is a risk that they will move in parallel.

As in the case of access to data and information, training and capacity building of the direct stakeholders can enable them to participate effectively in the interaction on an equal footing with other stakeholders by creating a level playing field. It should also be recognized that this might involve positive discrimination and corresponding weighting for the poor and the disadvantaged.

## 11.3 REFERENCES

Manasi, S. Raju, K.V. Latha, N. and Nagothu, U.S. (2008) Competing Water Uses in Tungabhadra Sub-basin, India. STRIVER Policy Brief No. 5, available at www.striver.no

Manasi, S., Latha, N., Paranjape, S., Joy, K.J., Nagothu, U.S., Raju, K.V. and Mollinga, P.P. (2009) Strategies and Recommendations towards an IWRM approach in Tungabhadra sub-basin, India. STRIVER Policy Brief No. 15.

Mollinga, P. (2006) IWRM in South Asia: A Concept Looking for a Constituency. In Mollinga P., Dixit, A. and Athukorala, K., *Integrated Water Resources Management: Global Theory, Emerging Practice and Local Needs*, Sage Publications, New Delhi.

Mollinga, P.P., Pillai, R., Saravanan S.V., Joy, K.J., Paranjape, S., Portela, M.M., de Almeida, A.B., Machado, M., Lana-Renault, N. and García-Ruiz, J.M. (2008) Strategies for rainfed and irrigated agriculture: Tungabhadra basin, India and Tagus basin, Spain and Portugal. STRIVER Task Report No. 9.2, available at www.striver.no

Nagothu, S.U. and Manasi, S. (2008a) IWRM and Livelihoods: Fisheries in Tungabhadra sub-basin, India, STRIVER Policy Brief No. 1, available at www.striver.no

Nagothu, U.S. and Manasi, S. (2008b) Fisheries in Tungabhadra Basin, India: Current Status and Future Possibilities to improve the livelihoods of marginal communities, STRIVER Report 9.5. p. 46, EU FP6 project. (www.striver.no)

Lo Porto, A., Barkved, L. and Gosain, A.K. (2008) Modelling water and nutrients balance in Tungabhadra river basin: scenario analysis and management recommendations, STRIVER Policy Brief, PB No. 7, available at http://kvina.niva.no/striver/Portals/0/documents/STRIVER_PB7.pdf

Paranjape, S. and Joy, K.J. (1995) Sustainable technology: making the Sardar Sarovar project viable – a comprehensive proposal to modify the project for greater equity and ecological sustainability, Ahmedabad: Centre for Environment Education.

Paranjape, S., Joy, K.J., Manasi, S. and Latha, N. (2008) India IWRM and traditional systems: tanks in the Tungabhadra system, Policy Brief No. 4, available at http://kvina.niva.no/striver/Portals/0/documents/STRIVER_PB7.pdf

Society for Promoting Participative Ecosystem Management and International Water Law Research Institute (2007) The Tungabhadra River basin proceedings of the first stakeholder meeting", held on 9–10 January 2007 at MYRADA Training center, Danapur, Hospet.

SOPPECOM (2008) Innovative technology and institutional options in rainfed and irrigated agriculture. STRIVER Summary Report No. 9.6, available at http://kvina.niva.no/striver/Portals/0/documents/STRIVER_D9_1Task_Report_9_6.pdf

Vaidyanathan, A. (ed.) (2001) Tanks of South India, Centre for Science and Environment, New Delhi.

# Chapter 12

# IWRM status in the Sesan River basin

G.D. Gooch and A. Rieu-Clarke

## 12.1 INTRODUCTION

This chapter describes the potential for IWRM and improved water governance in the Sesan River, which flows from the Central Highlands of Vietnam to the north-east provinces of Cambodia. The Sesan is characterized by the remoteness of the regions through which it flows and by the intensive development of hydro-electric power (HEP) in the upper, Vietnamese reaches of the river. While the development of HEP in Vietnam is considered by the Vietnamese to be necessary in order to meet the demands of increases in standards of living in their country without depending unnecessarily on the coal mined in the north-east of the country, the changes in river flows caused by the dams providing this energy have created major problems for the rural populations living along the river in Cambodia and have also negatively influenced fish production in the Sesan and perhaps even the Mekong itself. Besides the seasonal and temporal changes in the quantity of water flowing there is also evidence to indicate that water quality has been negatively influenced by the construction of the HEP dams in Vietnam (see Tiodolf, 2009 and Chapter 13).

From an IWRM point of view one of the main challenges in the Sesan is the increasing level of conflict between the interests of economic development and environmental protection. The regions, through which the Sesan flows, both in Vietnam and Cambodia, are characterized as areas with high original levels of biodiversity (Figures 12.1–12.2 and http://www.ibatforbusiness.org/) and at the same time widespread poverty. The high levels of biodiversity have, however, decreased in recent decades as a result of the intensified use of natural resources due to increasing populations and industrialized methods of production. Some of these important areas of conflict are around the conversion of forest areas into agricultural land (especially for industrial and commercial crops), the negative impacts of hydropower production on hydrological regimes, water quality, biodiversity and fishery, amongst others. From the transboundary point of view there are conflicts too, especially the conflict between upstream and downstream users (Gooch and Rieu-Clarke, 2010). In addition, the viewpoints surrounding water-use rights seem to be different in Vietnam and Cambodia. This is further compounded by a lack of exchange of information and communication between Vietnam and Cambodia (Gooch, 2008). The results of research conducted in the basin by Striver researchers indicates that the main transboundary problems in the Sesan basin are derived from the management of the hydropower reservoirs, located near the Vietnamese border with Cambodia. The main cross-border water resource issue is therefore the downstream impact of the hydropower plants on Cambodia, particularly the releases from the hydropower dams on the provinces of Ratanakiri and Stung Treng in north eastern Cambodia (Wyatt and Baird, 2007).

Stakeholder and public participation are central aspects of IWRN and as part of the STRIVER project, stakeholder workshops were organized in the Sesan basin (two in Vietnam and one in Laos), as well as field trips to Cambodia (Rieu-Clarke, *et al.*, 2010). The purpose of the workshops and field interviews was to bring together a range of stakeholders (civil society organisations, government – both national and provincial, international organizations, research institutions, etc), in order to exchange ideas on the key challenges facing the Sesan and propose alternative futures through the development of

 *Integrating Water Resources Management: Interdisciplinary Methodologies and Strategies in Practice.* Edited by Geoffrey D. Gooch, Alistair Rieu-Clarke and Per Stålnacke. ISBN: 9781843393252. Published by IWA Publishing, London, UK.

alternative scenarios (Gooch, *et al.*, 2010). A number of key challenges and opportunities to IWRM and water governance emerged from these stakeholder discussions; these will be discussed in more detail in this chapter.

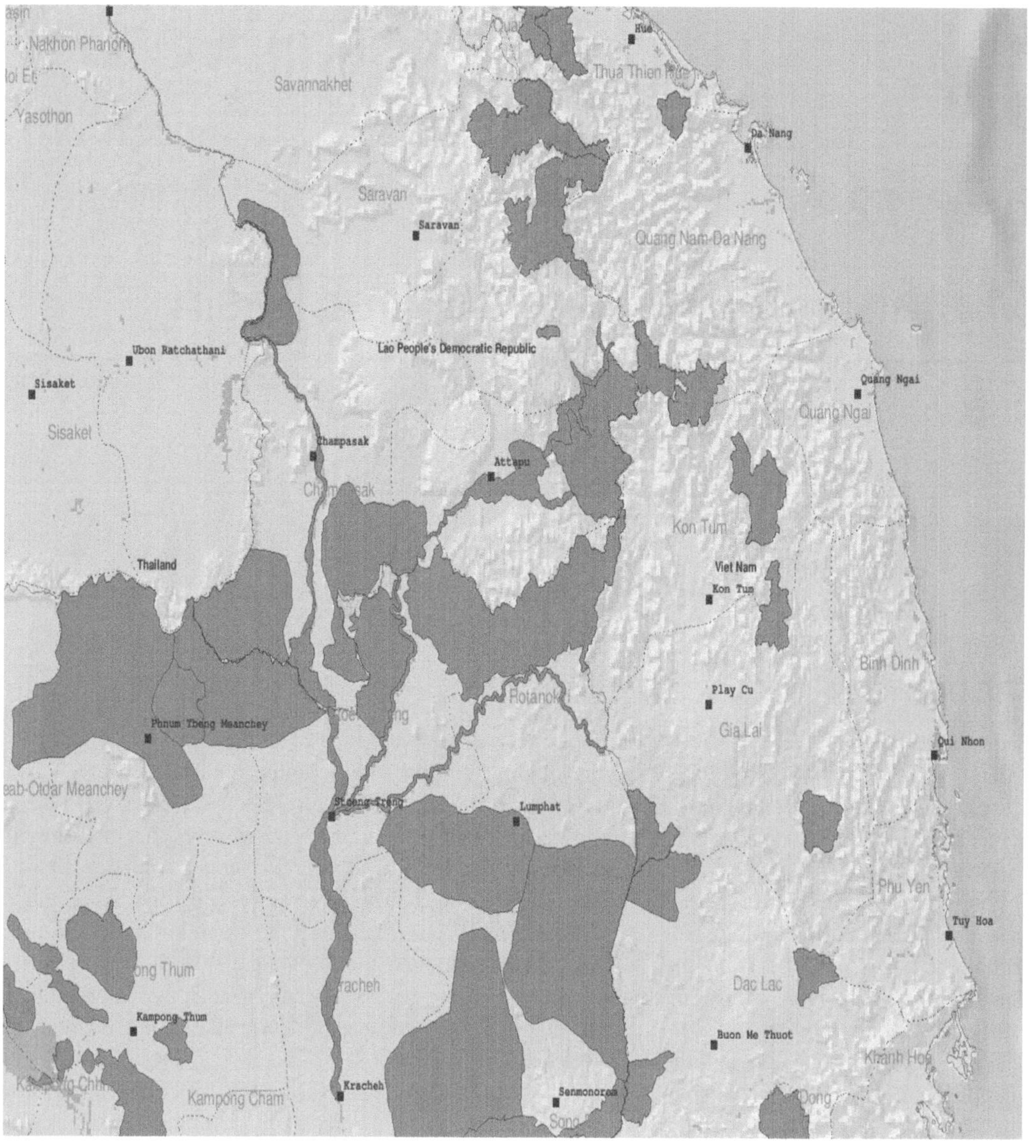


**Figure 12.1** Map of 3S Rivers basin, showing Key Biodiversity Areas (Source: www.ibatforbusiness.org).

In terms of the legal framework there was a common recognition among the people involved in these workshops and interviews that strengthening both bilateral and national laws would improve the implementation of IWRM principles. Most notable in this regard was the perceived need to harmonize national laws – particularly sectoral laws – and to develop supporting secondary legislation. At the bilateral level, the need to establish an appropriate agreement – under the umbrella of the 1995 Mekong Agreement – to promote the equitable and sustainable use of the entire Sesan basin was recognized, giving due recognition to the livelihoods of local communities. From an institutional standpoint, a strong recommendation was the need to improve coordination at the bilateral level through the establishment of a river basin organization. While positive steps have been taken through bilateral committees it was felt that such institutional arrangements need to be improved. Strengthening the institutional architecture was also considered to be a key challenge at the national and provincial levels, where a more decentralized community based approach was widely recognized as beneficial.

Communication and public participation were issues that were repeatedly raised by stakeholders. Most notable was the need to improve access to information, both between governments (international), between government departments (inter-sectoral), between national and provincial administrations, and between government and citizens. The need to increase participation of local community groups was also highlighted. The current national policies of both countries recognize the importance of stakeholder interaction and the policy interests need to be put in practice. In terms of capacity and training the stakeholders identified a number of areas that could be strengthened from the national governmental level to the community level, and ranging from enhancing management skills to training local communities in more efficient water use technologies. The priority is to build the capacity of managers and strengthen their competence in IWRM. A number of donor programmes already exist that need to be customized to the local needs.

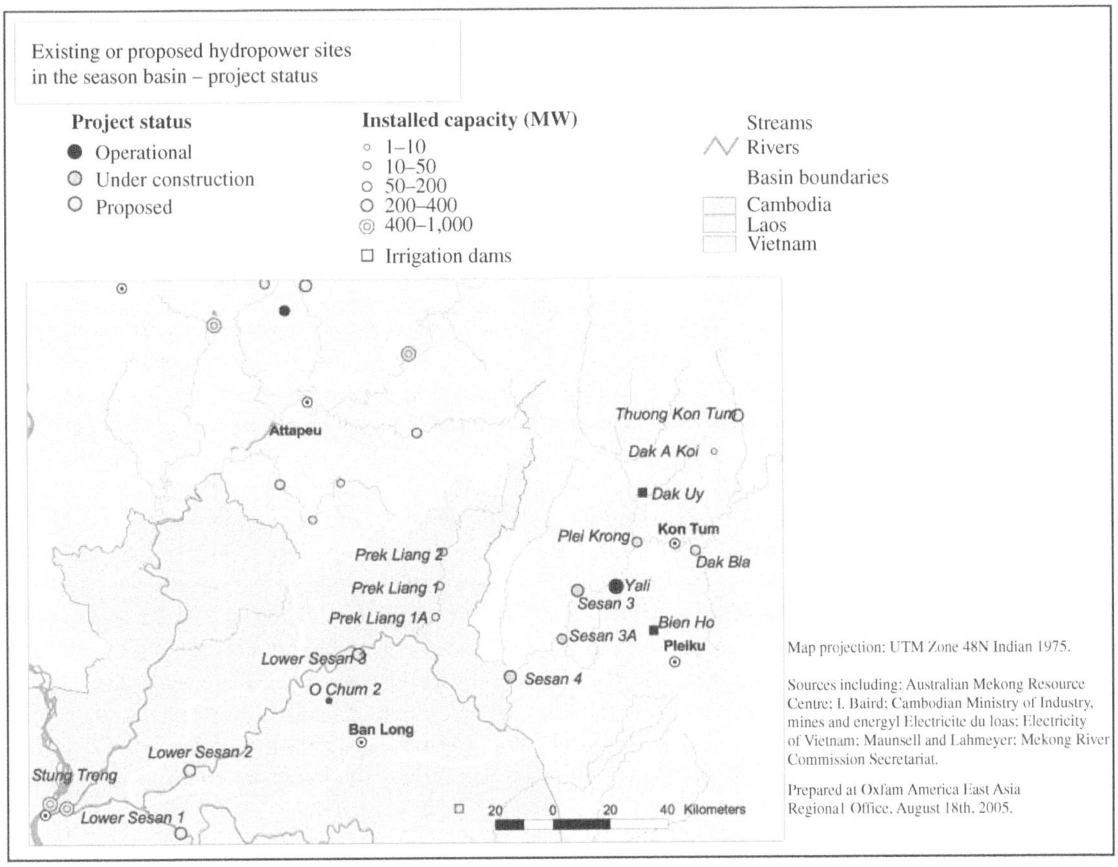


**Figure 12.2** Map of existing or proposed hydropower sites in the Sesan.

## 12.2 THE BASIN CONTEXT

The river originates in the mountainous areas of Vietnam's Dak Lak, Gia Lai and Kon Tum provinces before entering north-east Cambodia, where it moves down from the hills into relatively lowland areas. In Cambodia, the Sesan flows from the north-eastern border with Vietnam, west through Ratanakiri province and into Stung Treng province where it converges with two other rivers, the Sre Pok and the Sekong, before joining the Mekong. The Sesan and Sre Pock Rivers are together one of the largest tributaries of the Mekong River and provide around 10% of the water flowing in the lower reaches of the river (Halcrow and Partners, 1998). The drainage area of the river is about 18,600 km$^2$, with nearly 60% of this located in Vietnam and 40% located in Cambodia and the greatest part of the river is therefore in Cambodia. At the Vietnamese – Cambodian border the average flow over the year is about 380 m$^3$ per second. In Vietnam the Sesan has three main tributaries, the Krong Poko, Dak Bla and Sa Thay Rivers although there are a total of 83 tributaries that are more than 10 km long. The average slope of the basin in Vietnam is about 14% and there are numerous steep sections that create many waterfalls and rapids. More details concerning the basin characteristics are provided in the following Chapter.

Both Cambodia and Vietnam are party to numerous international and regional conventions related to water resources and the environment, including the Ramsar Convention on Wetlands (1971), the UN Convention on Biodiversity (1992), the UN Climate Change Convention (1992), and the UN Desertification Convention (1994). In the context of the management of water resources, both States are party to the 1995 Mekong Agreement (MRC, 1995). Article 1 of the Mekong Agreement stipulates that,

> "the parties agree to cooperate in all fields of sustainable development, utilization, management and conservation of the water and related resources of the Mekong River Basin including, but not limited to irrigation, hydro-power, navigation, flood control, fisheries, timber floating, recreation and tourism, in a manner to optimize the multiple-use and mutual benefits of all riparians and to minimize the harmful effects that results from the natural occurrences and man-made activities" (MRC, 1995).

## 12.3 IWRM STATUS IN THE SESAN

According to research conducted in the Striver project the implementation of this and other agreements is however problematic, and constitutes the main challenge facing IWRM and water governance in the Sesan (Rieu-Clarke and Gooch, 2010). In both Vietnam and Cambodia water related authorities/organizations exist at various levels i.e., at the national, province, district and commune levels. In Vietnam, these authorities/organizations are mostly controlled by the central state. In Cambodia the situation is quite different. Due to the large number of research institutions and multilateral and bilateral aid programmes in Cambodia and the Great Mekong sub region (GMS), as well as national and international NGO's, the Sesan represents a case where there are a multitude of actors, both at national and international levels.

Since the Sesan is a tributary to the Mekong River the Mekong River Commission (MRC) has negotiated a series of transboundary water allocation rules to achieve a reasonable and equitable utilization of the Mekong river waters, as envisioned in the 1995 Mekong Agreement. As part of the co-operation within the MRC, the Sesan hydropower development in Vietnam is now based on the Agreement on Co-operation for the Sustainable Development of the Mekong River Basin and associated rules and procedures, which represent an important legal framework for hydropower development in the river basin.

Cambodia and Vietnam are both members of the MRC and have also set up the Sesan River Committee for the prevention of any further negative impacts on the river. The two parties have also established mechanisms of regular information exchange, emergency alerts on the flow regime, operating process and discharge of water from the hydropower plants, as well as flood alerts and river water quality control. Most people living in the riparian areas in the catchment area belong to ethnic minorities. They are traditionally dependent on subsistence agriculture and fishing and have over time developed techniques suited for small-scale water utilization. Population density in the region is relatively low as a result of the relatively small share of agricultural land. Towns such as Pleiku and Kon Tum in Vietnam provide commercial centres for the predominantly agricultural populations living in the countryside and agriculture still represents more than half of total GDP, with services and industry on the increase.

Poverty and famine is still part of the realities of life for up to 10% of the population in Vietnam, and even higher, up to about 20%, in smaller villages and in the mountain areas and food shortages are not uncommon. Poverty is also the result of difficulties in getting access to health and education facilities in the more isolated villages (World Bank, 2010).The mainstay of the economy of the basin is agriculture and the share of industry in the economy is still small. Average income per capita is low, although it has increased from 139 USD in 2000 to 196 USD in 2005 in Gia Lai province of Vietnam. However, for standards of living to be increased, more improvements in infrastructure, education and health care facilities are necessary. Poverty levels are even higher in Cambodia.

The increase in population and the need for modernizing the societies have been the main drivers for the increased utilization of water resources, particularly with regard to hydropower production, large scale irrigation, water supply and also discharge off effluents. It is the central authorities that are primarily promoting large scale water resources utilization projects. Most of the positive impacts seem to have been at the national level, especially on national economic development, while for many rural people these projects have created unforeseen negative impacts. For instance, local inhabitants live according to the old or traditional modes of water use utilization which, by and large, cannot be integrated with the new forms of water resources utilization. The livelihood options of the local population living along a regulated river are, in most cases, also diminishing.

Though the Sesan River comes under the auspices of the MRC, this region lacks the tradition of civil society participation and the state remains the main decision-maker, with limited support for public participation in the formal legal framework. However, NGOs are creating space for stakeholders, especially in Cambodia, through the formation of NGO forums and through NGO participation in projects and networks. For instance, the World Conservation Union has both governmental and NGO members, as well as technical commissions and country and regional programmes. While some NGO activities are not integrated into the policy system, the expertise of the NGOs and the materials produced by them could be a resource for stakeholders.

## 12.4 ENVIRONMENTAL IMPACT ASSESSMENTS (EIA) AND THE HYDROPOWER DEVELOPMENTS IN THE SESAN

Throughout Vietnam a concerted effort has been made in recent years to systematically implement EIAs within all sectors, from the central administration to the local level. The last 10 years or so has also witnessed a significant increase in the total number of EIA reports and registration forms which have been considered and approved. However, the 2006 Environment Report highlights that there are still challenges for effective implementation of EIAs; such challenges include (a) the ratio between formal EIAs, and the total number of projects and units required to make EIAs; and (b) a variation in the standards of implementation within various regions and sectors (MoNRE, 2006). Similarly, the report notes that detailed regulations in support of EIAs are still lacking in Vietnam. The issue of EIAs is particularly prevalent in the context of the Sesan and hydropower development taking place within Vietnam (Wyatt and Baird, 2007). In 1993, shortly after EIAs became a legal condition in Vietnam, pursuant to the 1993 Environmental Law, an EIA for the Yali Falls dam was conducted by the Vietnamese state-owned electricity utility company, Electricity of Vietnam (EVN), and a Swiss based consultant, Electrowat Engineering Services Ltd. The EIA assessed downstream project impacts, but with a very limited reach. The study was funded by the Swedish aid agency, SIDA, and the Swiss government, and coordinated by the Interim Mekong Committee. Wyatt and Baird (1997) note:

> 'From the start, the EIA adopted a flawed, limited definition of the downstream area. Project impact boundaries were confined to upstream resettlement areas and the powerhouse. The project's downstream impact area was considered to extend only 6 km from the dam, with downstream impacts in Cambodia not even investigated or contemplated. The downstream population was identified as 'sparse' and the impact negligible'.

While it is true that rural populations by the Sesan in Cambodia are not large, they are still negatively affected by the HEP dams and their livelihoods have been compromised. The downstream impact area is obviously much larger than the 6 km downstream of the dams, and the EIA's failure to include Cambodia does not rhyme well with the ambitions of the MRC or prevailing international agreements. EVN defended their decision not to extend the reach of the EIA further downstream on the basis that the reach of the SeSan in Cambodia was inaccessible during the time due to security concerns. However, according to Hirsch, (2006) such an explanation does not explain why downstream impacts in the 80 km stretch of the Sesan from the Yali dam to the Cambodian border were not considered, or why studies were not carried out when the security situation improved in the mid-1990s. Similarly, an EIA for the Sesan 3 dam commissioned by EVN, funded by the Asian Development Bank and carried out by Worley WTL Ltd., failed to assess downstream impacts in Cambodia. Writing in 2007, Wyatt and Baird (2007) noted that,

> 'The EIAs for the Sesan 3 and Sesan 3A dams have not been officially released, and the Cambodian government has not received these studies. Yet construction of both dams is well under way. Construction on Sesan 3 began in June 2002, and work on Sesan 3A began in April 2003. To date, the Sesan 4 dam is the only one for which the Vietnamese have provided as copy of the EIA to the Cambodian government for comments. For that dam, they released an English-language version of the feasibility study and EIA in August 2005, and requested comments from the Cambodian government by December 2005. However, because construction actually began in January 2005, the purpose of seeking comments at such a late stage is questionable".

Subsequently, an EIA was commissioned to "evaluate the impacts in Cambodia from the development and operation of Ialy (Yali) Hydropower Project and from future hydropower development on the Se San River in Vietnam". SWECO in association with the Norwegian Institute for Water Resources, Enviro-Dev and ENS Consult conducted the EIA. The report was released in May 2007 (SWECO, 2006). A joint meeting of the Vietnamese and Cambodia National Mekong Committee was held in July 2007 to present the report. According to Probe International several Cambodian NGO's voiced their dissatisfaction about the time period from publication and public consultation, and that the report was not available in Khmer, the language used by many of the riverside communities (Probe International, 2007).

## 12.5 RECOMMENDATIONS FOR THE SESAN

### 12.5.1 Governance of the Sesan

The institutions of governance must be accountable to the citizens of a country in order for us to be able to talk about 'good governance'. With respect to accountability, both the Cambodian and Vietnamese states provide the right of access to justice through their Constitutions and secondary legislation and transparency should, officially, be guaranteed. However access to justice is also dependent on access to objective and unbiased legal representatives; the number of trained lawyers in these countries currently hinders such access. Also, as in many places in the world, difficulties remain in the implementation of water legislation. For IWRM and water governance to be successful, it is also vital that access to information is assured and that justice can be guaranteed through functioning legal and political systems. It is not enough, however, to simply provide information through the production of official documents etc.; this information must also be communicated in ways suitable for the designated audience and receivers.

IWRM and good governance are also dependent of stakeholder participation and in the case of the Mekong in general, and the Sesan in particularly, stakeholder participation in key decision making processes has unfortunately traditionally been lacking. However, a concerted effort has been made in recent years through the Mekong River Commission (MRC), and the national governments to engage with stakeholders in IWRM related issues. Finally, concerning IWRM, both countries have adopted new water laws that promote some of the key principles of IWRM (Law on Water Resources, 1998; Law on Water Resources Management, 2007). However, these water laws are in need of strengthening, particularly in relation to the adoption of detailed secondary legislation to support their implementation. Similarly, at the basin level and in co-ordination with the MRC, there is a need – and encouragingly also a recognition – to strengthen the governance arrangements needed to achieve the equitable and sustainable management within the entire Sesan River basin. In terms of implementation, there is a need to strengthen the system for civil society access to redress and remedy, and thus improve accountability. In Vietnam, while the 2002 Law on the Organization of People's Courts centralized the appointment of judges from the Local People's Council to the Chief Justice of the Supreme Courts; and the budgeting of local courts – previously decided by the provincial Departments of Justice – now rests with the National Assembly, there still remains concern over the independence of Vietnamese Courts (Global Integrity, 2006). In addition, the Supreme Court has only recently started publishing decisions. Similarly, a UNDP survey concluded that confidence in the court system in Vietnam is relatively low (UNDP, 2004). Vietnam ranks no. 121 out of 180 in the 2001 Transparency International Corruption Perceptions Index (Transparency International, 2008). Similarly, the National Integrity Systems Country Report for Cambodia notes that while the independence of the judiciary is officially assured, "practically this is not the case" because judges use the advice of the Ministry of Justice in interpreting often-vague legislation (Transparency International, 2006). Cambodia ranks no. 166 out of 180 (joint with Kyrgyzstan, Zimbabwe, Uzbekistan and Turkmenistan) in the 2008 Transparency International Corruption Perceptions Index (Transparency International, 2008).

As the Sesan is a transboundary river its governance is dependent on the cooperation of the two riparian states, Vietnam and Cambodia. Vietnam has constructed HEP stations on its side of the border in the upper reaches of the river, activities in Vietnam, which has had considerable effect on conditions in downstream Cambodia (Figure 12.3). Among the most important of these activities is the timing (daily or otherwise) of the releases of water from the dams, the volume and velocity of these flows, and also at which time of year large amounts of water are released (SWECO, 2006). Customary international law obliges states to notify each other of planned measures on international watercourses that may have a significant adverse effect in other states, and these conditions obviously apply to the Sesan. Provisions related to notification and consultation are included in the 1995 Mekong Agreement, and more detailed provisions for the Mekong have been developed through the Procedures for Notification, Prior Consultation and Agreement, which were approved by the Mekong Council in 2003. However, in relation to the Sesan, Wyatt and Baird (2007) observe that in the case of

> '...all of the Sesan dams, planning and construction have preceded any notification by the Vietnamese to the MRC and the Cambodian government. While the Agreement does not require consultation for intra-basin tributary use of water, the Vietnamese government did not consult the Cambodian government on the EIAs for the individual Sesan dams before construction' (Wyatt and Baird, 2007).

**Figure 12.3** Sesan downstream communities in Veun Sai, Ratanakiri province, Cambodia (Photograph: G.D. Gooch).

Concerning the exchange of data and information Article 24 of the Mekong Agreement provides that one of the functions of the Joint Committee is to regularly obtain, update and exchange information and data necessary to implement the agreement (MRC, 1995). Moreover, the MRC has established procedures for the exchange and sharing of data and information in 2001. In addition, a key objective of the Mekong Agreement, as articulated in Article 2, is

> '...to promote, support and coordinate in the development of the full potential of sustainable benefits to all riparian States and the prevention of wasteful use of Mekong River Basin waters, with emphasis and preference on joint and/or basin-wide development projects and basin programs through the formulation of a basin development plan, that would be used to identify, categorise and prioritise the projects and programs to seek assistance for and to implement at the basin level.'

Joint surveys, investigations and studies are therefore incumbent in the objectives of the Mekong Agreement, as well as the resultant functions of the Council, Joint Committee and Secretariat. However, in relation to public access to data and information there are no explicit provisions within the Mekong Agreement and it must also be remembered that the Mekong Agreement is dependent on the good-will of the signatory states. Never the less, an objective of the Procedures for Data and Information Exchange and Sharing is to, "make available, upon request, basic data and information for public access" (MRC, 2003). In addition, the MRC as custodian of information has established the MRC-Information System, designed to support planning, development, decision making and monitoring activities under the Mekong Agreement. Guidelines on the Custodianship and Management of the MRC Information have also been adopted, which *inter alia*, stipulate that the MRC, in line with established policies, will make appropriate data and information available to the public through various media (MRC, 2005). No special obligations and responsibilities apply to the general public other than those provided by any applicable laws, however, and in the case of the Sesan, many rural people living by the river in Cambodia are illiterate or unable to access and understand this kind of information.

At the national level, Article 69 of the Vietnamese Constitution provides citizens with a basic right to receive information (Constitution, 1992). Similarly, Article 4 of the Law on Freedom of the Press provides a Citizens' right to, "access via the press, information on all aspects of domestics and world affairs." More specifically, in relation to water, Decree 162/2003/ND-CP (2003) regulates the collection, management, exploitation and use of data and information on water resources. Article 104(2), of the Law of Environmental Protection (2005), also requires that, "[i]nformation must be made public in forms easily accessible by concerned organizations and individuals". However, such access is conditioned on Decision 212/2003/QD-TTg, (2003) which provides a list of State secrets classified as top secret in the field of natural resources and the environment, which includes information on water resources, geology and minerals, environment, hydrometeorology and surveying. The list of State secrets include, *inter alia*, (i) policies of and the state of disputes related to establishment, maintenance and development of international river basin organisations; and (ii) disputes relating to the exploitation, use and protection of international water resources. It should also be remembered that the Sesan originates in the Central Highlands of Vietnam, a region that has been exceptionally sensitive from the Vietnamese government's point of view as it experienced a period of unrest at the end of the 1900's and more significantly now, is a border region with a country with which Vietnam has not always had the best relations.

In Cambodia, Article 16 of the Law on Environmental Protection and Natural Resources Management (1996) provides that, 'the Ministry of Environment, following a request from the public, shall provide information on its activities and shall encourage public participation in environmental protection and natural resource management.' However, the procedures for such access to information should be more fully outlined in a sub-decree from the Ministry of the Environment and it is not clear if such a sub-decree has yet been issued. It should also be noted that information provision is acknowledged in the Governance Action Plan II 2005–8 (2005), in the context of access to judicial information, for example, and the environment. As noted earlier, however, it must be remembered that access to this type of information is dependent on the population's ability to obtain, read and understand it and this in often not the case for rural populations on the Sesan. The situation in Cambodia is also characterized by the influence of international and national NGO's and their important role in the spread of information. This information is in general very negative to the construction of HEP dams and stresses the detrimental effects on rural communities in north-east Cambodia. While there have undoubtedly been many negative influences on village life by the Sesan in Cambodia, not all are the result of Vietnam's HEP dam construction. Overfishing and the use of illegal fishing techniques such as hand grenades, electricity and cyanide have also played an important role.

The UN Common Country Assessment for Vietnam (2004) notes that, "access to information is a major constraint on public participation in decision-making. Information is sometimes shared through one-way communication channels, effectively restricting the ability of people to express their views and of local officials to benefit from useful information relevant to decision-making. However, in recent years the mass media have played an increasingly important role in disclosing incidences of corruption and abuse of power." In addition, the Country Assessment notes that,

> "...transparency remains an important issue at all levels of government. Some government agencies operate on the assumption that the public neither has the right to access information relating to the business of government nor the capacity

to use this information in socially constructive ways. More freedom to report corruption cases would discourage corrupt practices and increase public awareness of rights and laws" (UN Common County Assessment for Vietnam, 2004).

In 2006, Vietnam's Environmental Protection Agency, in coordination with the provinces produced and updated information for an electronic bulletin on the river basin environment, and some localities have developed an environment database for their own localities. However, as noted in the Environment Report (2006), "there are no databases at the basin level or sub-basin level. The exchange, sharing of environmental data and information among provinces within river basins and among basins is still very poor". In providing a broad overview of the situation in relation to water resources, the National Water Resources Strategy notes that,

"...the current extent and condition of national water resources has not been fully assessed, and the data on water resources is not reliable and comprehensive. The collection, management and archive of information on water resources are still dispersed. Data in different agencies is inconsistent and is not exchanged among state agencies. Data and necessary information on water resources, and changes in water resources are the basis for making plans for socio-economic development in provinces and territories, but are not adequate and are not regularly updated. Information management is not efficient and there is no national databank for water resources. Systems for reporting and providing data and information on water resources and on water exploitation and use of industries, sectors, organisations and individuals are not well focused" (Ministry of Natural Resources and Environment, 2006).

This situation, in which information is dispersed, has partly been a result of the competing competences of The Ministry of Agriculture and Rural development (MARD) and The Ministry of Natural resources and the Environment (MONRE), both of which have been responsible for different parts of water management (Gooch and Rieu-Clarke, 2010).

In the context of the Sesan River basin access to information on hydropower developments – and particularly their impacts in Cambodia – has been limited. The Sesan 3A HEP dam's EIA, for example, was drafted in April 2000, but not made public due to Vietnam refusing a request from the ADB to make it so (Wyatt and Baird, 2007). However, a range of data and information on natural resources, socio-economic conditions and sector development within the Sesan region is available. An ADB report (2008) observes that such information has been collected, "through the work of the BDP1 and the MWRAS-World Bank". The report goes on to note that

"...the scope of this information is generally national and provincial, but provides little or no data at the district level. Numerous other sources of data and reports are available in the Line Agencies either at national level or at local level, or at project and NGOs levels; e.g., a well-documented situation analysis has been carried out in the Stung Treng province. Sector databases are under development at MRCS-BDP level, one concerning the hydropower sector and one concerning the irrigation sector".

## 12.5.2 Participation

Stakeholder and public participation are cornerstones of IWRM and good water governance and have come more and more to be seen as necessary for effective decision-making and implementation. The Mekong Agreement (1995) however, does not provide for public participation within decision-making concerning international watercourses such as the Sesan River. Within Vietnam, pursuant to the Law on Environmental Protection (2005) participation of relevant public societies, organisations and individuals within SEA and EIA process is included. During the EIA process comments and feedback from commune level People's Committees and representatives of local communities within the project areas, as well as any opposition to the plans, should be incorporated into the EIA report. Then during the EIA appraisal stage, local organisations, communities and individuals are entitled and encouraged to provide comments to the relevant authorities, and the latter must take account of all comments before any decision is made. In addition, approval of the EIA must be reported to the local people's committee in the relevant area(s). Other legislation related to public participation includes the Government Decree 79/2003/ND-CP on Democracy at Grassroots Level, the 2003 Construction Law (details the collection of public opinion on master plans), and Decree no. 88/2003 provides a clear legal basis for the formation and management of non-governmental organisations (United Nations Country Team, 2004).

Despite this, in practice, participation in decision-making generally and specifically relating to water remains under-developed both within Vietnam and Cambodia. The Global Integrity Report (2006) observes that Vietnam represents a special case in the context, as the Communist Party dominates government and no independent civil society organisations exist. However, while traditionally top-down government structures sometimes act as an impediment to public participation, the formal structures of participation do provide certain channels for participation. Never-the-less, an ADB report related to water sector reforms, notes that the participation of civil.

Society and government-community partnerships remain weakly developed and that most water-related activities are conducted by the government. Such sentiment is reiterated by the Vietnamese National Water Resources Strategy, which observes that, "enhancing public awareness and consciousness in protecting and preserving the resource is still limited. The poor results in this hinder the participation of the entire society in the protection of water resources"(Ministry of Natural Resources and Environment, 2006).

A similar story can be found in Cambodia, although here the context is radically different. The United Nations Development Framework Report on Cambodia notes that

> "...civil society organizations and Cambodian citizens continue to face significant challenges, for example in accessing information, establishing dialogue and developing participatory processes. The Government's capacity to engage in policy dialogue with civil society is growing, but participatory mechanisms have yet to be regularized and mainstreamed into policy development, public planning or monitoring processes" (United Nations Development Framework, 1998).

The repeated exhortation to improve public participation is echoed in the NGO Papers on Cambodia's development (2006), which suggests that participation level are not as high as high as is wished. In relation to the Cambodian side of the Sesan and the difficulties faced by local communities due to the hydropower development upstream, the 3Ss Protection Network (3SPN) was formed in 2001. The objective of 3SPN is to "document the social, economic and environmental impacts of the dam and to serve as a legitimate representational body" Trandem (2008). While largely through informal channels, the umbrella organization claims that it acts as a network for dialogue and negotiation related to resolving dam-issues and ensuring the equitable use of water resources in the 3S region. In reflecting on the achievements of the network, Trandem (2008) observes that,

> "by challenging stakeholders to engage in dialogue, the network has raised awareness of the impacts suffered by communities downstream of Yali Falls dam and has forced many stakeholders in the Mekong River basin to reconsider the meaning of good water governance and recognize the need for greater stakeholder involvement in the decision-making processes related to water use in the Mekong and its tributaries. In demanding that stakeholders be held accountable and that affected communities have the ability to participate in decisions about their lives and futures, the network has endured both successes and challenges".

However, Hirsch and Jensen (2006) notes that, "to date, there has been little that could realistically be termed negotiation between the communities and 3SPN, on the one hand, and Electricity of Vietnam on the other. Governance arrangements entered into by the intermediaries in whom the communities placed some hope at the outset, notably the CNMC and MRCS, have not proven responsive."

Participatory poverty assessments indicate that women's participation is limited in commune and village level affairs in rural areas such as the Sesan basin in the Central Highlands. Contributing factors include time constraints as women continue to shoulder the bulk of domestic work while also working outside the home, lower educational attainment and disempowerment. The situation is similar in Cambodia. Only five ministries were noted by NGOs as having "Gender mainstreaming policies", but only four of these had actually carried this policy into implementation according to the NGO Papers on Cambodia's development (2006). A total of ten ministries were under an obligation to put such plans in place.

## 12.6 INSTITUTIONAL CHALLENGES RELATED TO IWRM IN THE SESAN

In Vietnam past overlaps have led to a re-assignment of roles and responsibilities amongst institutions involved in water resources management. As noted above, during 2007, MoNRE has taken over most responsibility for water resources management, and the Vietnam Mekong River Committee and Water Resources Management Agency (WRMA) are situated within the Ministry. At the political level the Provincial People's Committees are important, as are the different governmental departments at the provincial level. When resolving issue relating to water resources use in a river basin WRMA invites both representatives of the People's Committee and the Departments in the province. The People's Committee is the governing authority, and they formulate policy which should then be implemented by the departments; in this respect not only DoNRE are involved, but also other related departments in the province such as Department of Investment and Planning (DARD), Department of Traffic, etc. River basin planning organisations are being development in the Vietnamese part of the lower Mekong River, and for the Dong Nai and Thai Binh river basins, yet these RBOs are not as yet particularly active, and there are problems related to weak institutional responsibility and poor coordination. For example, in Vietnam the water resources management implementation is mainly under the responsibility of the Provincial Peoples Committees, while at the same time national (non-provincial) institutional such as Electricity of Vietnam (EVN) play an important role in issues devoted to water use in terms of hydropower development. Moreover, the Ministry of Agriculture and Rural Development (MARD) is the chair of the RBOs while at the same time has limited influence on water resources issues as this is under the governance of the province administration. Furthermore, the various RBO offices are placed in the different institutions under MARD with the main role as technical assistance rather than overall management states at the state and provincial levels. The provincial authorities do not consider these offices as the water resource management institution. However, new legal frameworks and initiatives taken in 2006, such as that of the Cau River (decision 174/2006/QD-TTg, 2006) may help remedy these problems, by demonstrating that management plans that include economic aspects as well as acceptance of the importance of sustainable development and protection of the environment can be formulated.

Sub-basin level planning was also been highlighted as an issue during the Sesan Stakeholder Workshops (Figure 12.4), while the main river was managed by the government, there is no planning for the tributaries. According to development plans for the Kon Tum province, 30 small hydropower stations (5–50 MW) are planned for the tributaries of the Se San. Greater consideration therefore needed to be paid to small to medium scale hydropower development and irrigation systems in the tributaries of the Se San and their impact on the basin as a whole. Similarly it can be noted that at the provincial level there are therefore four organizations which are of equal role for IWRM. These are DoNRE (WRM in general, licensing authority, water resource monitoring), DARD (water supply and irrigation works), the Department of Planning and Investment (appraising and financing projects) and the Department of Industry (hydropower projects). Besides these, there are some related organizations such as the Department of Science and Technology, the Department of Health, etc. There are also organizations such as the Irrigation Division and the Centre for Freshwater and Sanitation both of which belong to DARD. In the DoNRE there are some divisions related to WRM, such as Division of Natural Resources (water resources management), the Division of Environment (EIA, water pollution) and the Division of Land Use and Planning (land management). At the District level the most important are the Department of Economy which belongs to District People's Committee (water supply and irrigation works management), the Irrigation Work Exploiting Company which belongs to DARD, the Water Use Complex in the district, communes (which exploit small irrigation works), the irrigation works construction companies (many private companies) and the hydropower project construction companies (joint-stock, private companies for small hydropower projects).

**Figure 12.4** 3rd STRIVER Sesan Stakeholder workshops, Vientiane December 2008 (Photograph: G.D. Gooch).

The Environment Report (2006) notes that, "the coordination of ministerial, sectoral and local offices for resolving the problems of the river basin is still weak. The various localities in the river basin have no common voice, unity and close cooperation in the environmental management of the river basin. The concepts of responsibility from the localities and sectors on environmental protection in the river basin are still inadequate. Some localities have wrong concepts of environmental protection purposes for the river basin and their roles and responsibilities in organizing environmental protection in the river basin." Similar sentiments were voiced during the 1st Sesan STRIVER Stakeholder Workshop, where the group recognized a clear need to improve the management of the Sesan and its tributaries. There was considerable overlap of water management competencies between various departments, e.g., environment, fisheries, agriculture and between various tiers of government, e.g., district, provincial and national. It was felt that the main ministries and their provincial departments responsible for water did not cooperative effectively. There was a need for comprehensive and integrated planning for water resources – combining land, forest and water resources – in order to allocate and share benefits and impacts among the different sectors. The group emphasised the need for stakeholders and the authorities to cooperation, which should encompass cooperation with key actors within the Cambodian section of the Sesan.

Within the Cambodian part of the Sesan, the Provincial Governments and Governors of Ratanakiri and Stung Treng have played an active role in investigating and voicing concerns over the developments within the Basin. The Ratanakiri

Governor was the first to announce the problems along the Sesan to the media. Both Ratanikiri and Strung Treng governments have supported the findings of impact studies on the Yali, and appealed to the national governments to find a solution to the issues. In publicly released statements at the National Se San Workshop, the Rataniakiri and Stung Treng governments have supported the findings of the Ratanakiri and Stung Treng impact studies, arguing that the reported impacts were not present before Yali was built. Cautiously they have supported the view that the wide range of impacts is partly caused by the operations of Yali Dam and that this was causing economic hardship for communities.

The role of the MRC within the Sesan has also been raised as an issue since the impact of the Yali Falls dam has become widely known. When directly questioned in 2002 about the role of the MRC, the CEO pointed to that fact that Yali preceded the formation of the MRC in 1995, and that the matter should be taken up directly with the Cambodian government (Wyatt and Baird 2007). However, as noted previously, the MRC during its 7th Council Meeting in October 2000 facilitated the establishment of the Cambodia-Vietnam Joint Committee for the Management of the Sesan River. The MRC has provided facilitation, language interpretation, and technical information and advice for these meetings.

## 12.7 CONCLUSIONS

IWRM in the Sesan is as yet under-developed, partly due to the economic difficulties experienced in the basin, especially the Cambodian part of it, and partly because of the need to produce energy for rapid modernization and poverty alleviation in Vietnam. These aims are at present leading that country to prioritize HEP production over environmental concerns. Looking at the five aspects of IWRM assessment outlined in Chapter 1, it is possible to draw a number of conclusions concerning IWRM and the Sesan River. The aspects are:

(1) protection of the catchment and the environment;
(2) measures to ensure efficient and equitable use of water;
(3) institutional analysis and stakeholder involvement in water management;
(4) capacity-building; and
(5) transboundary issues.

Taking the aspects one by one it can be said that in the case of protection of the catchment and the environment (1), there are major challenges in the Sesan. The HEP dams built in the Central Highlands of Vietnam have radically changed the natural water flows of the river and have disrupted the seasonable and daily changes in water levels. These changes have had serious negative impacts on the ecology of the river, with adverse effects on the spawning grounds of fish as well as on the vegetation in the riparian zones that is dependent on certain levels of water at certain times of the year. They have also had major influences on the possibilities of local, rural populations to continue their traditional ways of life based on river-side farming. Alco in the case of to ensure efficient and equitable use of water (2) there are significant problems. The use of water is now dominated by the needs of the upstream user, Vietnam, which manipulates water flows according to its needs to produce electricity. The needs of the down-stream country, Cambodia, are not being taken into account to any large degree. The institutional analysis and examination of stakeholder involvement in water management (3) has also shown that the institutions that are in place such as the MRC, have been unable to manage the conflict of interest that arose. This is partly because the MRC is a forum for discussion without a clear mandate to govern the tributaries of the Mekong (Rieu-Clarke and Gooch, 2010). The real power of the MRC lies in the hands of the member states national governments, working through their national Mekong River Committees; although the influence of the Asian Development Bank in promoting participation has also been notable in recent times. Capacity building (4) is another major challenge to IWRM in the Sesan, especially in the case of Cambodia. The Cambodia authorities suffer from a lack of resources and education, according to the water and fisheries managers that were interviewed in the Striver project, and in some cases there was simply no funding for the transport necessary to carry officials to more out of the way villages along the river. Local rural populations are also handicapped in their efforts to understand developments and influence their futures by illiteracy and very low levels of education. If capacity building was to be seriously implementedit should involve the education of officials as well as villagers; it should also involve increased funding in order to enable monitoring of water levels, water quality and fishing. The final aspect outlined above, that of transboundary issues, (5) is of course central in the case of the Sesan. The transboundary nature of the challenge facing IWRM has been noted in Chapter 1, so it will suffice here to state that the problems facing transboundary management are substantial; at the same time there are indications that the need for a transboundary commission is now realized, and there is therefore hope that in the future the management of the Sesan River will be the responsibility of a combined Vietnamese-Cambodian institution.

## 12.8 REFERENCES

Asian Development Bank, *Sesan, Sre Pok and Sekong River Basin Development Study in Kingdom of Cambodia, Lao People's Democratic Republic, and Socialist Republic of Viet Nam*, Draft Inception Report, June 2008.

Biodiversity Convention (1992): UN Convention on Biological Diversity, 5 Jun. 1992 (entered into force 29 Dec. 1992), *International Legal Materials* **31**, 822.

Climate Change Convention (1992) UN Framework Convention on Climate Change, 9 May 1992 (entered into force 24 Mar. 1994), *International Legal Materials* **31**, 849.

Constitution (1992) Constitution of the Socialist Republic of Vietnam (as amended 25th Dec., 2001), http://www.vietnam laws.com/.

Decision 212/2003/QĐ-TTg (2003) on the List of State Secrets classified as Top Secret in the Field of Natural Resources And Environment, Oct. 21, 2003, http://vbqppl2.moh.gov.vn/law/.

Decision No. 174/2006/QD-TTg (2006) *Approving the Master Plan on the Protection and Sustainable Development of the Landscape and Ecological Environment of Cau River Basin.*

Decree 162/2003/ND-CP (2003) (Vietnam) on Regulating the Collection, Management, Exploitation and Use of Data and Information on Water Resources.

Desertification Convention (1994) Convention to Combat Desertification in those Countries Experiencing Serious.

Drought and/or Desertification, Particularly in Africa, 14 Oct. 1994 (entered into force 26 Dec. 1996), *United Nations Treaty Series* **1954**, 3.

Global Integrity (2006) *2006 Country Reports – Vietnam*, http://www.globalintegrity.org/reports/2006/vietnam/index.cfm/

Gooch, G.D. (2008) http://kvina.niva.no/striver/Portals/0/documents/STRIVER_PB6.pdf Communication in IWRM in Transboundary Rivers. STRIVER Policy Brief No. 6, http://kvina.niva.no/striver/Disseminationofresults/STRIVERPolicyBriefs/tabid/78/Default.aspx

Gooch, G.D. and Rieu-Clarke, A. (2010) The Science-Policy-Stakeholder Interface and Transboundary Regimes. In *Science, Policy and Stakeholders in Water Management*. Gooch, G.D. and P. Stålnacke (eds.), Earthscan, London.

Gooch, G.D., Allan, A., Rieu-Clarke, A. and Baggett, S. (2010) The Science-Policy-Stakeholder Interface in Sustainable Water Management. Creating interactive, participatory scenarios together with stakeholders. In *Science, Policy and Stakeholders in Water Management*. Gooch, G.D. and P. Stålnacke (eds.), Earthscan, London.

Governance Action Plan II 2005–8 (2005) *Accelerating State Reform. Strengthening Governance and Improving Public Services Delivery, Supreme Council for State Reform*, www.car.gov.kh

Halcrow and Partners (1998) *Sekong-Se San and Nam Theun River Basins Hydropower Development Study*. Manila, Asian Development Bank.

Hirsch, P. and Jensen, M.K. (2006) *National Interest and Transboundary Water Governance in the Mekong*, http://www.mekong.es.usyd.edu.au/projects/mekong_water_governance.htm/

Law on Water Resources (Vietnam), No. 8/1998/QH10, May 20, 1998.

Law on Water Resources Management of the Kingdom of Cambodia, 29 June 2007, http://www.mowram.gov.kh

Law of Environmental Protection (2005) (Vietnam), Dec. 12, 2005, http://www.cleanairnet.org/caiasia/1412/articles-72598_resource_1.pdf/

Law on Environmental Protection and Natural Resources Management (1996) (Cambodia) http://cambodialaw.ifrance.com/envirnt/tocenvgb.htm

Ministry of Natural Resources and Environment (2006) *National Water Resources Strategy* Culture-Information MoNRE (Ministry of Natural Resources and Environment) (2006) *Environment Report of Vietnam 2006 – The Current State of Water Environment in 3 River Basins of Cau Nbue-Day and Dong Nai River Systems*, http://www.nea.gov.vn/pdf_soe2006_eng/phandau.pdf

MRC (Mekong River Commission) (1995) Agreement on the Cooperation for the Sustainable Development of the Mekong River Basin, http://www.mrcmekong.org/agreement_95

MRC (Mekong River Commission) (2003) Procedures for Notification, Prior Consultation and Agreement, http://www.mrcmekong.org/download/agreement95/Procedures_Guidelines/Procedures-Notification-Prior-Consultatio-Agreement.pdf

NGO Papers on Cambodia's development (2006) *Monitoring of Joint Monitoring Indicators and Implementation of the National Strategic Development Pla*n, http://www.ngoforum.org.kh/Development/Docs/doc.htm

Probe International, for the NGO Forum on Cambodia, *Review of Electricity of Vietnam's Environmental Impact Assessment (EIA) on the Cambodian Part of the Se San River Due to Hydropower Development in Vietnam* (2007), http://www.probeinternational.org/files/pdfs/muw/SeSanEIA_Review_Probeinternational.pdf

Ramsar Convention (1971) Convention on Wetlands of International Importance especially as Waterfowl Habitat, 2 Feb. 1971 (entered into force 21 Dec. 1975), **996** *United Nations Treaty Series* 245.

Rieu-Clarke, A. and Gooch, G.D. (2010) Governing the Tributaries of the Mekong – The Contribution of International Law and Institutions to Enhancing Equitable Cooperation over the Sesan. *Pacific McGeorge Global Business and Development Law Journal* **22**(2), 193–224.

Rieu-Clarke, A., Baggett, S., Campbell, D., Joy, K.J. and Paranjape, S. (2010) The Science-Policy-Stakeholder Interface in Sustainable Water Management: Creating Interactive Participatory Scenarios together with Stakeholders. In *Science, Policy and Stakeholders in Water Management – An Integrated Approach to River Basin Management.* Gooch, G.D. and Stålnacke, P. (eds.) (2010) Earthscan, London.

SWECO Groener, *Final Report – Environmental Impact Assessment on the Cambodian Part of the Se San River Due to Hydropower Development in Vietnam* (2006).

Trandem, A. (2008) A Vietnamese/Cambodian Transboundary Dialogue: Impacts of dams on the Se San River, *Development* **51**, 108.

Transparency International (2006) *National Integrity Systems Country Report: Cambodia*, http://www.transparency.org

Transparency International (2008) *Transparency International Corruption Perceptions Index*, http://www.transparency.org/news_room/in_focus/2008/cpi2008/cpi_2008_table

Tiodolf, A.M. (2009) A limnological study of the Sesan River in Cambodia in the dry season: focus on toxic cyanobacteria and coliform bacteria. Masters thesis. Ås: Norwegian University of Life Sciences. Department of Plant and Environmental Sciences, 106.

United Nations Country Team, *United Nations Common Country Assessment for Viet Nam* (2004) http://www.undp.org/rbap/Country_Office/CCA/Cca-Vietnam2004.pdf/

United National Development Programme (2004) *Access to Justice in Vietnam*. United Nations Development Programme, Hanoi 2004, http://www.undp.org.vn/undpLive/System/Publications/Publication-Details?contentId = 1824&languageId = 1./

United Nations Development Framework (1998) Cambodia, http://www.undg.org/unct.cfm?module = CoordinationProfile&page = Country&CountryID = CMB&fuseaction = UN%20Country%20Coordination%20Profile

Wyatt, A.G. and Baird, I.G. (2007) Transboundary impact assessment in the Sesan River Basin: The case of the Yali Falls Dam, *International Journal of Water Resources Development* **23**(3), 427–442.

# Chapter 13

# Water quality in the Sesan River – Role of natural scientific facts and figures

*P. Stålnacke, A.M. Tiodolf and D. Berge*

## 13.1 INTRODUCTION AND STUDY OBJECTIVE

The very first high-level policy summit of the Mekong River Commission (MRC) was held very recently in April 2010. The summit gathered high-level political leaders from the various countries within the basin, in addition to a wide range of experts in the field of integrated water resources management. The final declaration – also entitled as the 'MRC Hua Hin Declaration' – was signed by the heads of the governments of the Kingdom of Cambodia, the Lao People's Democratic Republic, the Kingdom of Thailand, and the Socialist Republic of Vietnam (MRC, 2010b). In the declaration, nine focus and priority areas was identified by these high-level government officials. Five of these nine topics -directly quoted from the summit declaration – were:

- *"Adopting and implementing the IWRM-based Basin Development Strategy";*
- *"Intensifying efforts to effectively manage the risks from flood, drought and sea level rise including establishment of forecasting and warning systems across the whole basin";*
- *"Monitoring and taking measures to improve water quality in priority areas of the Basin";*
- *"Identifying and advising on the opportunities and challenges of hydropower and other infrastructure development in the Basin, especially risks as they pertain to the protection of food security and livelihoods";*
- *"Continuing to improve the implementation of the Procedures for Data and Information Exchange and Sharing, the Procedures for Water Use Monitoring, the Procedures for Notification, Prior Consultation and Agreement and the Procedures for Maintenance of Flows on the Mainstream and finalize the Procedures for Water Quality";*

Our interpretation of these five priority areas, (which also reflect the content and spirit of this book), is the presence of increased political and management interest at high levels and subsequently a need for an integrated science-based study devoted to natural scientific monitoring data and water quality in a IWRM context. It implies to us that access to current and accurate water resources information throughout a basin like Sesan is emphasized as an emerging need and that monitoring of hydro-meteorological conditions (rainfall, waterflows etc.) and water quality need to be shared, as a prerequisite for building efficient cooperation in a transboundary setting.

In this chapter we discuss and also show-case the crucial role of water quality monitoring and sound natural scientific data, using the Sesan River as the case study. The river is both the major drinking water supply and livelihood source for the local people living along the river. In 1996, Vietnam started an expansion program of hydropower plants along the Sesan River system (see Chapter 12). There have been claims by various NGOs and official organizations that the water quality in the Sesan River has seriously deteriorated after the hydropower projects started. According to these sources of

 *Integrating Water Resources Management: Interdisciplinary Methodologies and Strategies in Practice.* Edited by Geoffrey D. Gooch, Alistair Rieu-Clarke and Per Stålnacke. ISBN: 9781843393252. Published by IWA Publishing, London, UK.

information, water quality problems originated from the Yali reservoir, which may be contaminated with toxic cyanobacteria (also called blue green algae), or possibly other toxic elements. The river has according to local people and various NGOs become turbid and produced a bad odor. Irregular water flow, increased erosion, decreased fish supply and polluted water are other claimed effects. At the same time there have been other contrasting but prevailing views aired. For example, Long Saravuth, deputy director of the Ministry of Water Resources' hydrology department, have said that a year-long study of monthly water from the river showed that it was safe. (The Cambodia Daily, Apr. 11, 2006). A study by the Mekong River Commission (MRC) based on available chemistry data in Sesan revealed that the river is classed as high-quality (MRC, 2008). Moreover, a recently published study from the Mekong River Commission – based on a bio-monitoring survey in Sesan – showed that the two sites monitored were in the 'excellent status' or 'good' status class and with no sign of change over the years covered, i.e., 2005–2008 (MRC, 2010a). Most notably, few other water quality surveys have been conducted in Sesan (The Fisheries Office and NTFP Project, 2000; Wyatt and Baird, 2007; Sweco-Grøner, 2007).

There is therefore a high need for further investigations of the water quality in the Sesan River. The objective of this study was to analyze the water quality in the Sesan River during the dry season, focusing on the presence of cyanobacteria, cyanotoxins and thermo-tolerant coliform bacteria, especially the indicator organism *E.coli.*

The limitation of our study is that it was conducted during a short monitoring period (two weeks duration) although the results are discussed in relation to the previous knowledge status. Thus, the quantitative results should not be interpreted in absolute terms, but more as an example of what water quality data can provide in terms of increased knowledge in a science-policy/management and IWRM setting. The final section discusses these aspects.

## 13.2 MAJOR FACTS ABOUT THE SESAN RIVER BASIN

The Sesan River is one of the largest tributaries of the Mekong River. It is a trans-boundary river originating in Vietnam and running through north eastern Cambodia, where it joins the Mekong River (Figure 13.1). The Sesan River has a drainage area of 18,600 km$^2$, with nearly 40% located in Cambodia and 60% located in Vietnam. The river flows through mountainous areas in Vietnam's Dak Lak, Gia Lai and Kon Tum provinces, before entering north east Cambodia, where it moves into relatively lowland areas. In Cambodia, the Sesan winds from east to west through Ratanakiri province and into Stung Treng province, out into the Mekong River. There are considerable differences in rainfall patterns during the year. The rainy season lasts from August through to November, with river flow normally peaking in September-October. Precipitation varies from approximately 1000 mm per year in the lowlands of Cambodia to 2200 mm per year in the highlands of Vietnam. At the border with Cambodia the average flow is about 380 m$^3$/s. The specific runoff of the Sesan river basin varies from around 40 l/s per sq km in the upper western part of the basin to 30–35 l/s per sq km in the lower parts.

**Figure 13.1** The transboundary Sesan river between Vietnam and Cambodia. (Berge and Nhung, 2008).

In Vietnam the Sesan river has three main tributaries, the Krong Poko, Dak Bla and Sa Thay rivers. There are a total of 83 tributaries (more than 10 km long). The density of rivers and streams is low, about 0.38 km/km$^2$ on average. The average slope of the basin is about 14%. In the river, there are numerous steep sections that create many waterfalls and rapids.

Most people living in the riparian areas belong to ethnic minorities and are involved in subsistence agriculture and fishing and have developed techniques suited for small-scale water utilization. Population density is relatively low, reflecting a relatively small share of agricultural land in the region. Agriculture still represents more than half of total GDP, with services and industry on the increase. Poverty and famine is still part of the realities of life for up to 10% of the population, and even higher, up to about 20%, in smaller villages and in the mountain areas. Poverty is also the result of difficulties in getting access to health and education facilities in the more isolated villages.

The mainstay of the economy of the basin is agriculture and the share of industry in the economy is still small. Average income per capita is low, having increased from 139 USD in 2000 to 196 USD in 2005 in Gia Lai province. To promote economic and social development in the region, the first priority ought to be given to better infrastructure, and also to improved education and health care facilities. Further work is also needed to alleviate poverty, and to ensure community services and income improvements that will enable villagers to participate in a modern economy.

The increase in population and the need for modernizing the societies have been the main drivers for the increased utilization of water resources, particularly with regard to hydropower production, large scale irrigation, water supply and also discharge of effluents. The main transboundary problems in the Sesan basin are derived from the management of the hydropower reservoirs, located at the Vietnamese border with Cambodia. The main cross-border water resource issue is the downstream impact of the hydropower plants on Vietnam, particularly the releases from hydropower dams, in Ratanakiri and Stung Treng provinces in north eastern Cambodia. The hydropower and management issues are further discussed in Chapter 12 and issues around environmental flows in Sesan can be found in Berge and Nhung (2008).

### 13.2.1 Example of official monitoring of water quality in Sesan

In Sesan there are three official water quality monitoring sites at Pleiku (Vietnam), Adong Meas (Cambodia) and Phum Pi (Cambodia (Figure 13.2). The field visit in March 2008 revealed that the monitoring site at Adong Meas seemed more or less abandoned with e.g., the meter sticks over-grown with bushes and other vegetation (Figure 13.3). The stations are monitored and analysis performed by national laboratories under the overall guidance of the Mekong River Commission within the Water Quality Network Programme which maintains a quality assurance program. Stations are sampled 12 times per year, usually on 15th of each month, at a depth of 0.5 m in the middle cross-sectional profile (thalweg), or at the point of maximal flow if the mid point is not representative.

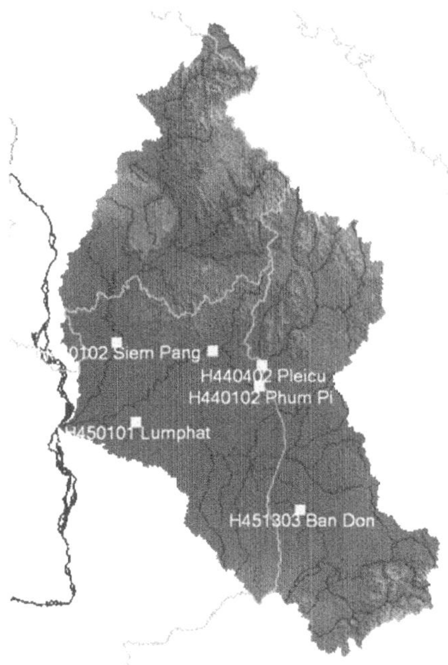


**Figure 13.2** Map of the Primary MRCs Water Quality Stations in the 3Ss Basins (Sesan, Srepok and Sekong): http://reta.3sbasin.org/

Bio-monitoring (i.e., benthic diatoms, zooplankton, littoral and benthic macroinvertebrates) has also been conducted in campaign measurements during the years 2005–2008 at two sites in Sesan (MRC, 2010a; see Section 13.1).

**Figure 13.3** The sampling site and meter stick at Adong Meas (site #2) overgrown with vegetation (Photograph: D. Berge).

## 13.3 FIELD STUDY AND METHOD

Our field visit was conducted during a two-week period in March 2008. Water samples were taken every second day in two different areas along the Sesan River; Adong Meas, 30 kilometers downstream the Vietnamese border, and Veun Sai, about 130 kilometers downstream the Vietnamese border (Figure 13.5). Water samples were also taken from three Cambodian tributaries of the Sesan River (O Kup, National Park Vireak Chey and O Tang). Each study area was visited three times. Average water temperature was 26,7 C and the average pH was 7,2.

Water for analysis was taken either with a water-sampler, or directly in a plastic container from the surface water. In addition, algae-samples were taken with a 25 μm net haul (Figure 13.4). Quantitative algae-samples and net haul samples were preserved by Lugol's solution (fytofix) for species composition analysis. The microscopy of phytoplankton was carried out in a laboratory in Norway. For analyzing cyanotoxins, 48 mm GF/C Wathman-filters were filtered with a known amount of water and dried in darkness. To detect the toxin-concentration the filters were analyzed at a laboratory in Norway by an anti-adda ELISA kit. To estimate the concentration of coliform bacteria 'Tubetest' were used. To detect the concentration of the indicator bacteria *E.coli* water-samples were cultured by membrane filter method in a field laboratory where selective medium, MLGA and VRBA, were used. A known amount of water was filtered through a 0,45 μm cellulose membrane filter and cultivated on the selected medium for up to five days in an incubator-oven at 44°C, distributed by 'Colifast'.

**Figure 13.4** Sampling of algae and foam floating on the Sesan River (Photograph: A.M. Tiodolf and P. Stålnacke).

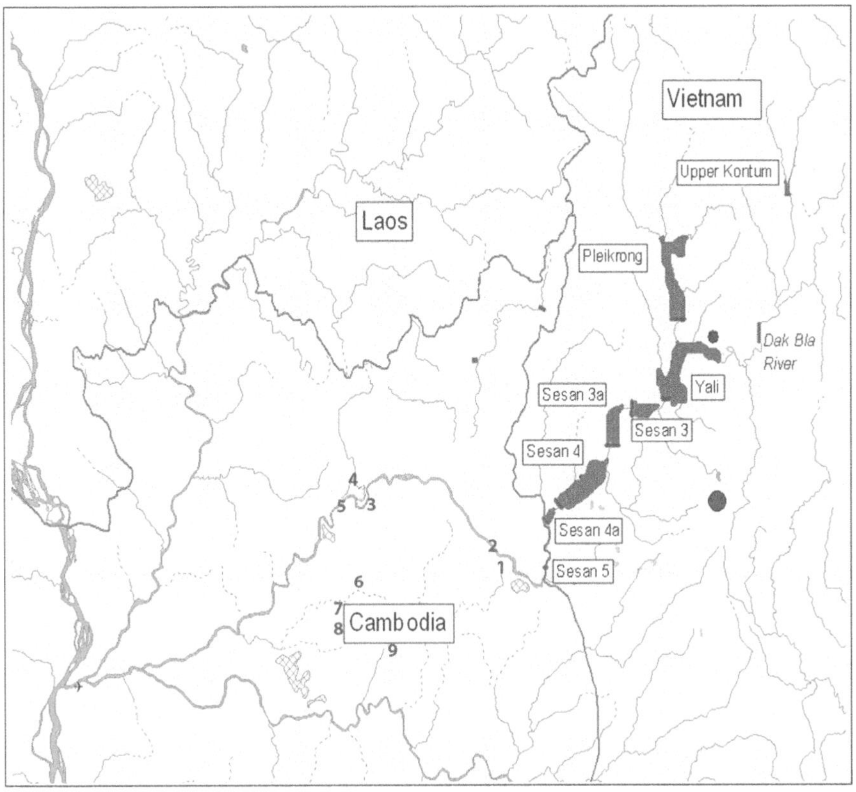


**Figure 13.5** Map over the study area. 1. The tributary O Kup, 2. Adong Meas, 3. Kachon 4. The tributary Vireak Chey National Park, 5. Veun Sai, 6. The tributary O Tang. 7–8. minor streams in the Ban Lung area and number 9. is the lake Kan San in Ban Lung. Source: Tiodolf (2009).

## 13.4 RESULTS

### 13.4.1 Health effects of cyanobacteria

Running river water is not a natural habitat for planktonic cyanobacteria. Occurrence of these bacteria in rivers is usually found in more stagnant water upstream. It is well-known that algal blooms sometimes occur in lakes and reservoirs with eutrophic conditions. Worthy of note here is that the scientific knowledge about eutrophication, in the vast majority, comes from northern temperate lakes in North America and Europe. Investigations in tropical systems are more scarce. Some examples of studied reservoirs with algal blooms in the tropics are the Lago do Paranoá in Brazil (Cronberg, 1978), Lake Kariba in Zimbabwe (Cronberg, 1997) and Tri An Reservoir in Vietnam. Tri An Reservoir is a multipurpose reservoir for, amongst other things, hydropower, irrigation, fisheries and supplies water for millions of people in Dong Nai and Binh Duong Provinces and Ho Chi Minh City (Dao *et al.*, 2010).

Some aquatic species of cyanobacteria can, under certain conditions, produce cyanotoxins which can be very poisonous and thus potentially dangerous to humans and animals. The most commonly registered symptoms shown by people exposed to cyanotoxins are allergic reactions i.e., fever, skin irritation and diarrhea. Microcystin is a hepatoxin, which means it is toxic to the liver and can cause injury to or cancer in the liver after a longperiod of exposure. Through interviews with the inhabitants along the Sesan River we learnt that irritation of the skin, gastric disorders and respiratory problems are commonly occurring after bathing in the river, especially amongst children and during the dry season (Figure 13.6).

There are several examples of human and animal deaths related to cyanotoxins. A well-known case is when over 50 patients died in 1996 in a hemodialysis center in Caruaru, Brazil (Komarek *et al.*, 2001). These patients had syndromes associated with liver hemorrhage or liver failure. Several different cyanobacteria, as well as microcystin, were detected in the water supply for the town Caruaru, in the reservoir Tabocas.

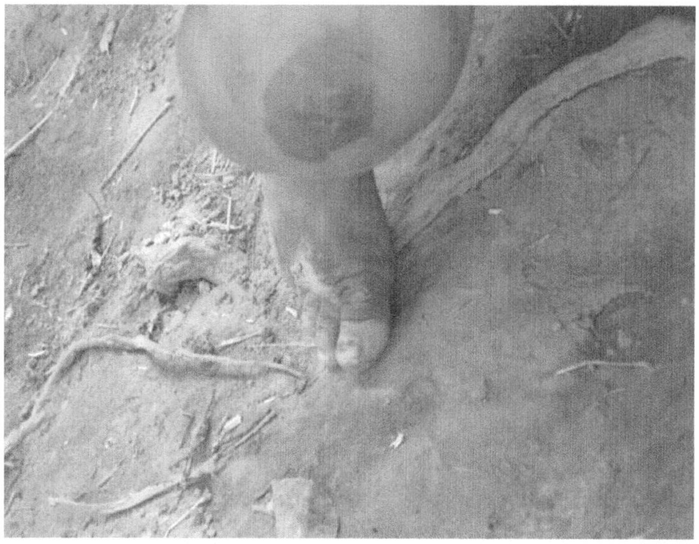

**Figure 13.6** A child with skin-problem living in Adong Meas (site #2) claimed to be caused by contact with the Sesan River water (Photograph: A.M. Tiodolf).

Cyanobacteria do not only produce hepatoxins. Like coliform-bacteria they are also gram- negative and can produce endotoxins from their outer cell membrane containing lipopolysaccharides (LPS) (Forssblad and Annadotter, 2008). According to their study, inhalation of aerosols containing endotoxins can cause an influenza-like reaction with fever but also respiratory problems, chock and irregular heart beating. The symptoms often appear one to six hours after exposure to endotoxins and in most cases last for approximately 24 hours.

In recent years, many studies have indicated a close relationship between blooms of cyanobacteria and following outbreaks of cholera. The theory is that the gel surrounding the cyanobacteria act like a reservoir, providing optimal nutrient conditions for the bacteria causing cholera, *Vibrio cholerae* (Ahmed *et al.*, 2008).

## 13.4.2 Cyanobacteria, toxins and algae composition in Sesan River

Existence of cyanobacteria in the Sesan River during the dry season was confirmed during our study (Figure 13.7). More specifically, in some more scientific terms, colonies of *Microcystis* spp. were identified in the quantitative analysis of water samples. In the qualitative analysis of net haul samples, colonies of *Microcystis* sp. as well as cellular fibers of *Planktothrix* sp. and *Aphanizomenon* sp. were confirmed. Colonies of *Pseudoanabena* sp. were established in cultures of water samples with Blue-Green agar.

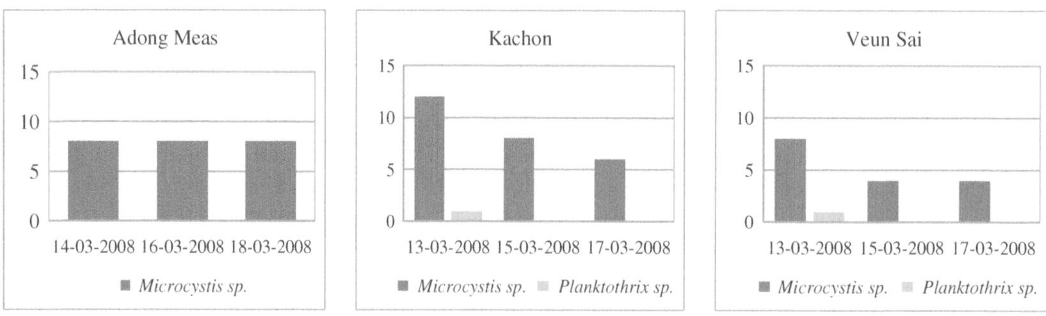


**Figure 13.7** Average density of cyanobacteria in the qualitative analysis of net haul samples from Adong Meas (site #2), Kachon (site #3) and Veun Sai (site #5). NB! A maximum value of 32 means a total dominance of a particular algae in the sample. Source: Tiodolf (2009).

In the Sesan River concentrations of microcystin (hepatoxin) (Table 13.1) exceeded the WHO limit for drinking water (1 μg $L^{-1}$) and approach the limit for safe bathing (10 μg $L^{-1}$). These limits are based on a critical level for an adult person with 60 kg weight exposure cyanotoxins.

Table 13.1 Concentrations of the cyanotoxin microcystin in water samples analyzed by an anti-adda ELISA kit (ABRAXIS, product no. 520011) at the NIVA-laboratory (Norway). Source: Tiodolf (2009)

| Locality (site #) | Date March 2008 | Filtrated amount of water (L) | Concentration of Microcystin (μg $L^{-1}$) |
|---|---|---|---|
| **Tributary** | | | |
| O Kup (1) | 14 | 1,27/1,05 | */* |
| O Kup (1) | 16 | 1,00/1,08 | */* |
| O Kup (1) | 18 | 1,00/1,00 | */* |
| **Se San** | | | |
| Adong Meas (2) | 14 | 1,20/1,30 | 5,23/8,39 |
| Adong Meas (2) | 16 | 1,62/1,88 | */6,96 |
| Adong Meas (2) | 18 | 1,00/1,00 | */* |
| **Se San** | | | |
| Kachon (3) | 13 | 1,00/1,00 | */* |
| Kachon (3) | 15 | 1,17/1,52 | */5,58 |
| Kachon (3) | 17 | 1,00/1,00 | */* |
| **Tributary** | | | |
| Vireak Chey National Park (4) | 15 | 0,75 (?)/0,75 | */* |
| Vireak Chey National Park (4) | 17 | 0,75 | * |
| **Se San** | | | |
| Veun Sai (5) | 13 | 0,73 | * |
| Veun Sai (5) | 15 | 1,00/1,05 | */* |
| Veun Sai (5) | 17 | 0,85/0,85 | * |
| **Tributary** | | | |
| O Tang (6) | 13 | 1,00 | * |

* Concentrations below LOD (limit of detection)

Both the algae and chlorophyll concentrations were different between the main stream and the local tributaries (Figure 13.8). Moreover, the algae composition in the analyzed water samples and net haul samples from the main Sesan River show predominance of the algae groups Dinophyceae, Chrysophyceae and Bacillariophyceae (Figure 13.9). Many of the specific species within these groups indicate a eutrophic environment. Whereas the samples from the local tributary rivers showed dominance of Chrysophyceae and μ-algae.

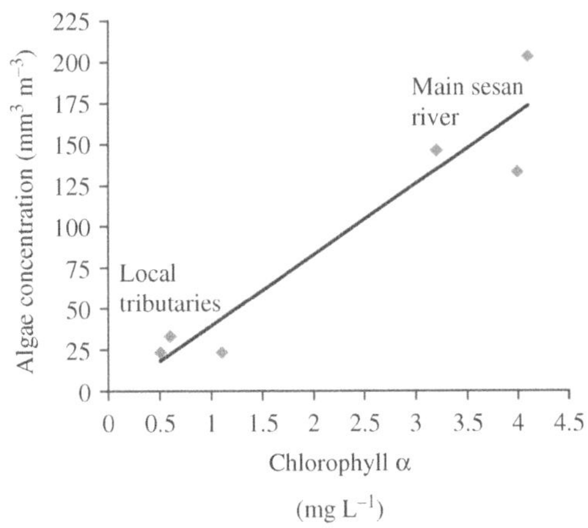


**Figure 13.8** Difference in the algae and chlorophyll a concentrations in water samples from the Sesan River (main stream) compared to water samples from the three tributaries (March 2008). Source: Tiodolf (2009).

Also noteworthy was that cyanobacteria made up a maximum of 3% of the phytoplankton composition in the quantitative samples from the main stream (i.e., water samples from Adong Meas) while no cyanobacteria were detected in the local tributaries.

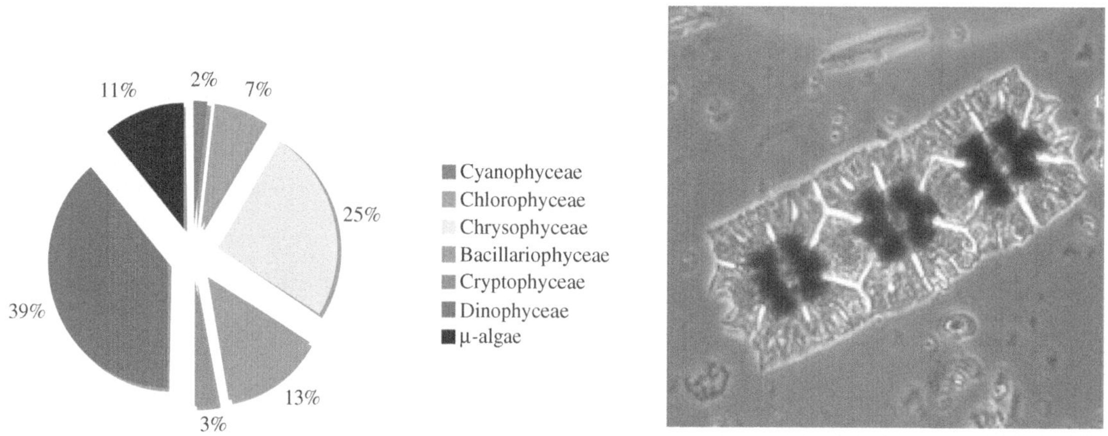


**Figure 13.9** Left panel: Average phytoplankton composition from all the samples in the main river of Sesan River. Source Tiodolf, 2009. Right panel: Chlorophyceae (green algae) (*Micrasterias foliacea*) from the tributary Vireak Chey National Park (site #4; see figure 13.5) (Photograph: R. Ptacnik, NIVA).

### 13.4.3 Thermo-tolerant coliform bacteria

The concentration of thermo-tolerant coliform bacteria was high; the indicator organism *E.coli* being $10^3$–$10^6$ CFU $ml^{-1}$ (Table 13.2). This indicates risk of spreading of waterborne diseases. WHO's limit for *E.coli* or thermo- tolerant coliform bacteria is that they should not occur in a 100 ml water sample. Similar levels of bacteria concentrations have been found through analysis of water samples in other tropical areas, for example Uganda, South Africa and India. Analysis of river water in Lebanon has shown concentrations of up to $6 \times 10^4$ CFU $ml^{-1}$ combined with significant health effects on people bathing in the river. Bathers got skin reactions, which is also one of the main health problems that the locals along the Sesan River experience in connection with bathing (see Figure 13.6).

Table 13.2 Water samples cultivated on the selective medium MLGA. Green colonies are *E.coli* and yellow colonies are thermo-tolerant coliform bacteria, which are not *E.coli.* Source: Tiodolf (2009).

| Locality | Date March 2008 | Cultivating Days 42–44° C | Thermo-tolerant yellow colonies (CFU 100 $ml^{-1}$) | Coliform green colonies (CFU 100 $ml^{-1}$) | Bacteria total (CFU 100 $ml^{-1}$) |
|---|---|---|---|---|---|
| **Tributary** | | | | | |
| O Kup (1) | 14 | 3 | * | $33\times10^5$ | $33\times10^5$ |
| O Kup (1) | 16 | 5 | * | $20\times10^5$ | $20\times10^5$ |
| O Kup (1) | 18 | 1 | | | ## |
| **Se San** | | | | | |
| Adong Meas (2) | 14 | 3# | * | $10\times10^6$ | $10\times10^6$ |
| Adong Meas (2) | 16 | 5 | * | $17\times10^5$ | $17\times10^5$ |
| Adong Meas (2) | 18 | 2 | $40\times10^5$ | * | $40 \times10^5$ |
| **Se San** | | | | | |
| Kachon (3) | 13 | 2,5# | $21\times10^4$ | * | $21\times10^4$ |
| Kachon (3) | 15 | 2 | * | $21\times10^4$ | $21\times10^4$ |
| Kachon (3) | 17 | 4 | | | ## |
| **Tributary** | | | | | |
| Vireak Chey National park (4) | 15 | 2 | * | $70\times10^4$ | $70\times10^4$ |
| Vireak Chey National park (4) | 17 | 2 | $9\times10^5$ | $8\times10^5$ | $17\times10^5$ |
| **Se San** | | | | | |
| Veun Sai (5) | 13 | 2# | $40\times10^4$ | * | $40\times10^4$ |
| Veun Sai (5) | 15 | 2 | * | $20\times10^5$ | $20\times10^5$ |
| Veun Sai (5) | 17 | 2 | | | ## |
| **Tributary** | | | | | |
| O Tang (6) | 13 | 2,5# | $85\times10^4$ | * | $85\times10^4$ |
| **Other watersystems in the Ban Lung area** | | | | | |
| Stream water (8) | 17 | 1 | $14\times10^4$ | $1\times10^4$ | $15\times10^4$ |
| Kan San (9) | 16 | 5 | $90\times10^4$ | * | $90\times10^4$ |

* No growth of thermo-tolerant coliform bacteria.
# Samples cultivated in 25–35°C.
## Not countable cultures due to too high concentrations of bacterial colonies.

### 13.4.5 Chemistry

Chemical water analysis show that the Sesan Rivers nutrient level is oligotrophic during the dry period. However, the water samples from the tributaries O Kup and O Tang show higher phosphorous concentrations and in some cases indicate eutrophic levels have been reached. The results are not shown here, but can be found in Tiodolf (2009) and Tiodolf and Stålnacke (2009).

## 13.5 FIELD STUDY SUMMARY

The following conclusions can be drawn from the sampling in March 2008:

- Occurrence of cyanobacteria and cyanotoxins in the main stream of the Sesan River was confirmed.
- No cyanobacteria in the water samples taken in three different tributaries were detected.
- The phytoplankton composition in the main stream of the Sesan River indicate a mesotrophic to eutrophic environment. The reason for this could be that the algae originate from more stagnant water, with higher nutrient concentrations, i.e., originating further upstream in the Sesan River.
- In contrast to the phytoplankton composition results, the water chemistry analysis in the main stream of Sesan River show an oligotrophic trophic level.
- According to the WHO limits, the results of this study show the drinking quality of the water in the Sesan River is not suitable for human consumption without treatment. More specifically, the levels of the cyanotoxin microcystin along with the content of thermo-tolerant coliform bacteria and the indicator organism *E.coli* were found to exceed the WHO safety standards which indicate risk of waterborne diseases.

## 13.6 CONCLUDING REMARKS

The important role of good water quality measurements is indisputable. This is not only a natural scientific dogma but also largely recognized by various international water management bodies. For example, the theme of the World Water Day in 2010 was in fact 'water quality'. On this day the International Water Association stressed "the urgent need to step up research, monitoring and assessment of water quality at global, regional, and local levels". The role of research results in sound policy formation and implementation was also high-lighted as another core message. Our scientific study in Sesan, despite being limited in duration, clearly shows that there is a high need for further monitoring. More specifically, our study indicates that the water quality in Sesan might be poor, which is somewhat contradictory to the far more optimistic views expressed by other sources such as the MRC and high officials (see Section 13.1).

## 13.7 REFERENCES


Ahmed, M.S., Hiller, S. and Lukas, B. (2008) Microcystis aeruginosa bloom and the occurrence of microcystin (hepapeptides hepatotoxins) from an aquaculture pond in Gazipur, Bangladesh. *Turkish Journal of Fisheries and Aquatic Sciences* **8**(1), 37–41.

Berge, D. and Nhung, D.K. (2008) The Use of Environmental Flows in IWRM with reference to the hydropower regulated Glomma River (Norway) and Sesan River (Vietnam and Cambodia). STRIVER Report D 8.1. Oslo.

Cronberg, G. (1978) The Lago do Paranoá Restoration Project. Phytoplankton ecology and taxonomy. (Final report). - PAHO/WHO, 77/WT/BRA/2341/04. (Mimiographed). *Limnologiska Institutionen*, Lund. pp. 71.

Cronberg, G. (1997) Phytoplankton in Lake Kariba, 1986–1990. In *Advances in Ecology of Lake Kariba*. Ed. J. Moreux,. Zimbabwe University Press, 66–119.

Dao, T.D. do-Hong, L.-C. Cronberg, G. Nimptsch, J. and Wiegand, C. (2010) Toxic cyanobacteria from Tri An Reservoir, Vietnam. *Nova Hedwigia* **90**(3–4), 433–448.

Forssblad, J. and Annadotter, H. (2008) Endotoxins in Swedish tap water, Swedish Water, Swedish Water Development, Report Nr. 2008–20. Stockholm: The Swedish Water & Wastewater Association (SWWA). 80 p.

Komarek, J., Azevedo, S., Domingos, P., Komarkova, J. and Tichy, M. (2001) Background of the Caruaru tragedy; a case taxonomic study of toxic cyanobacteria. *Algological Studies*, **103** (Cyanobacterial Research 2), 9–29.

Liljeström, I. (2007) Nitrogen and phosphorus dynamics in the Mekong basin: Nutrient balance assessment in a catchment scale. Master thesis. Helsingfors: Helsinki University of Technology, Department of Civil and Environmental Engineering. 116 s.

MRC (2008) An assessment of water quality in the Lower Mekong Basin. MRC Technical Paper No.19. Mekong River Commission, Vientiane. 70 pp.

MRC (2010a) Report on the 2008 biomonitoring survey of the lower Mekong River and selected tributaries. MRC Technical Paper No. 27. Mekong River Commission, Vientiane. 69p.

MRC (2010b) First MRC Summit & International Conference 2–5 April 2010, Hua Hin, Thailand. http://www.mrcsummit2010.org/index.html

SWECO Grøner (2007) Rapid Environmental Impact Assessment on the Cambodian part of the Se San River due to Hydropower Development in Vietnam. Final Report. Lysaker: SWECO Grøner. 204 s.

The Fisheries Office and NTFP Project (2000) A Study of the Downstream Impacts of the Yali Falls Dam in the Se San River Basin in Ratanakiri Province, Northeast Cambodia. Ban Lung, Ratanakiri Province, Cambodia: The Fisheries Office. 33 s.

Tiodolf, A.M. (2009) A limnological study of the Sesan River in Cambodia in the dry season: focus on toxic cyanobacteria and coliform bacteria. Master thesis. Ås: Norwegian University of Life Sciences. Department of Plant and Environmental Sciences. 106 s.

Tiodolf, A.M. and Stålnacke, P. (2009) A limnological study of the Sesan River in Cambodia in the dry season: focus on toxic cyanobacteria and coliform bacteria. STRIVER Technical Brief No. 12. www.striver.no

Wyatt, A.B. and Baird, I.G. (2007) Transboundary impact assessment in the Sesan River Basin: The case of the Yali Falls Dam. *International Journal of Water Resources Development* **23**(3), 427–442.

# Chapter 14

# STRIVER – Overall findings

*P. Stålnacke, G.D. Gooch and A. Rieu-Clarke*

## 14.1 INTRODUCTION

This final chapter of the book summarizes and discusses the main findings from the four Striver case basins; the Glomma (Norway), the Tagus (Spain and Portugal), the Tungabhadra (India) and the Sesan (Vietnam and Cambodia). Particular emphasis is placed on assessing the current status of IWRM and specific IWRM problems occurring in the basins; these include transboundary water governance, environmental flows, water pollution, land and water use interactions, stakeholder participation and basic environmental data. The conclusions of the experiences of the interactions between STRIVER scientists and local stakeholders are also presented in the Chapter.

In the following sections, the major findings from the project are summarized starting with a brief analysis of the IWRM status in the four basins followed by a summary of the specific IWRM issues dealt with in the project. These include:

(a) Scenario development;
(b) Stakeholder participation;
(c) Environmental flow methodology with science-stakeholder interaction;
(d) Transboundary conflicts and water governance;
(e) Pollution modeling with stakeholders;
(f) Land and water use interactions.

The chapter concludes with a set of policy guidelines derived from the different analyses conducted in the project and proposes initiatives concerning IWRM implementation in the case basins and beyond.

## 14.2 THE IWRM STATUS IN THE FOUR STRIVER BASINS

A comparison of IWRM status assessment of the four basins was undertaken as part of the project (Nesheim *et al.*, 2008) with a focus on environmental, socio-economic and institutional dimensions in the case areas. In addition, the potential for capacity building and the role of transboundary issues were analyzed as they were considered important for the STRIVER project and for implementing IWRM. The following major conclusions were drawn from this analysis.

Socio-economic development was the common driving force in all the STRIVER river basins, although the role played by each individual sector varied according to the local conditions. As a consequence, in some cases the river basins face similar pressures and comparable impacts; in some cases they were different. Negative changes in water quantity and quality were the two common factors in all the conflicts and all the river basins had developed plans or policies to attempt to protect most of the environmental aspects. However, it was difficult to assess their level of implementation and the effectiveness of the planned measures. The extent to which problems of water scarcity and water pollution are effectively mitigated depends not only on the existence of relevant law and policy, but also on the degree to which such instruments are implemented. In other words, it is the management and the institutional situation which

 *Integrating Water Resources Management: Interdisciplinary Methodologies and Strategies in Practice.* Edited by Geoffrey D. Gooch, Alistair Rieu-Clarke and Per Stålnacke. ISBN: 9781843393252. Published by IWA Publishing, London, UK.

ultimately determines the outcome. The case basin situations show that there are numerous issues of pollution, environmental flow, water allocation, and more, where laws and policies are not always implemented. This situation may be partly due to weak institutions (particularly in the Sesan and Tungabhadra), but also due to poor coordination between institutions. The latter problem is evident in almost all basins, caused by the multitude of different bodies involved, and unclear definition of roles and functions as well as (in some cases) a lack of financial capacity. In Sesan and Tungabhadra basins, insufficient monitoring (e.g., crucial water quality parameters) and control and also poor implementation of the prescribed actions were registered. Generally it can be argued that in areas where water shortage was evident, measures were taken for water saving and reuse.

In all the STRIVER basins, access to information and decision-making was present in varying degrees, in particular through environmental impact assessment procedures. Laws stipulating the right to information should also allow stakeholders access information through public authorities, although once again, this is not always the case in practice. Whilst a number of conventions and national laws and policy provide for stakeholder participation in water management, practice demonstrates that government officials, hydropower, industry and urban interests tend to dominate decision-making. As a result interest groups representing environmental concerns and communities tend to have less influence on decision-making, despite the procedural rules allowing them access to relevant information and procedures such as public inquiries.

Capacity building was officially emphasized as part of many water policies and strategy documents in the Sesan, Tungabhadra and Tagus case study basins. However, it was evident that these official statements were seldom operationalized to any larger degree by the authorities. Competence building which includes NGOs is increasingly becoming an integral part of basins and covers information campaign brochures and posters (Figure 14.1) for the public and training programs for certain sectoral groups (especially for local communities, farmers). In Cambodia, foreign donors were an important actor, responsible for training programs.

**Figure 14.1** Education poster for local farm-land fish-ponds in Cambodia. (Photograph: G. D. Gooch)

The case basins represent a full spectrum of alternative transboundary contexts/situations, from inter-states (Tungabhadra) to international (Tagus and Sesan). The cases illustrate that the extent of conflict decreases as we move along the continuum from international to local and the extent of conflict varies by type: e.g., competition over the quantity of water was more controversial than conflict over quality. Hydropower development is perhaps the most important issue in a transboundary context, as it may potentially have detrimental effects on the river downstream of the border. It can be seen that there exist a considerable number of 'IWRM-initiatives', IWRM-plans, and policies in all the four basins but practical full IWRM implementation is generally lacking, except in the case of Glomma.

## 14.3 SCENARIOS DEVELOPMENT

Scenarios can be used as a means of both improving policy strategies and of involving stakeholders and the public in that policy making (Gooch and Stålnacke, 2006). Scenarios present a way for the policy-maker or manager to test ideas about possible futures through exercises that can clarify the probable results of certain courses of action and implementation of particular governance frameworks. They can also be used as a means of improving public and stakeholder participation. If these groups are involved in the formulation and evaluation of scenarios, they can provide insights not readily available for policy-makers, as well as increasing the level of social learning.

Scenarios describe plausible futures; however, the utility of both the process and the products are also vested in the present and can also help to identify contemporary problems and issues. Knowledge about key drivers and uncertainties of the future can provide information towards better decisions in the present and a better understanding of key drivers and trajectories of changes will not only clarify the impact of decisions, but may facilitate active countering of undesirable trajectories of change. The identification and characterization of key uncertainties can enable a more structured approach to risk management. Strategies and decisions can be played out in different futures to secure the most beneficial outcome through the most robust approaches with the least risk. The process can produce new knowledge that can not only benefit resource managers and decision makers, but empower stakeholders. This extends from politicians and policy makers, through to government officials, private sector and civil society. Knowledge generated and sourced through structured participative research processes can increase understanding, both through the act of participating in the process and by accessing appropriate forms of communication. The identification, description and ranking of key drivers and uncertainties can be translated to potential implications. Scenarios have been the focus of increased interest and initiatives such as those published by the European Environmental Agency on the Environmental Scenarios Information Portal (http://scenarios.ew.eea.europa.eu/) have contributed to developments in this field, yet the use of scenarios as a means of improving stakeholder participation and the science-policy interface still needs to be developed.

Gooch *et al.* (2010) have earlier demonstrated how scenarios can be used as a means of two-way communication with stakeholders and civil society. The stakeholder groups formulated future possible developments for the project team, which then used them in their analyses. The meeting participants then identified who they saw as the main actors and problems. These were then incorporated, together with the results of the analyses of actor-networks, communication and legal aspects, into draft scenarios, which were discussed at a second stakeholder workshop. Following this, the scenarios were revised once again and finally presented at a third stakeholder workshop. Through this process, stakeholder perspectives, governance analyses and the results of scientific research in the river basins were combined. The stakeholder workshops also helped identify central aspects of the project such as the actor-networks. Finally, the development of scenarios highlighted particular aspects of the legal regime that may need to be altered and those aspects of the existing regimes that obstruct, affect or are likely to lead to particular scenarios, or that have an effect on the extent to which particular stakeholder groups can influence resource management.

The use of scenarios showed that they can be a useful tool to involve stakeholders and to improve the legitimacy and quality of policy strategies. As the meetings were held on both sides of the Tagus and Sesan rivers, stakeholders were able to formulate national perspectives which were then combined at meeting with all participants. In the case of the Sesan river this meeting, held at the MRC in Vientiane in December 2008, was in itself an achievement, as a problem which has complicated the formulation of policies for the sustainable use of the Sesan River has been a relative lack of communication between the Vietnamese and Cambodian stakeholders, especially at the provincial level.

The major challenges to the use of scenarios are the following:

- The process is time-consuming, and in order to succeed it is necessary to organize at least three meetings; one to introduce the idea of scenarios, which can be difficult to grasp at once, a second meeting to discuss the first draft versions, and a third to finalize the scenarios.
- The selection of participants for the meetings is a central issue; it is at the same time necessary to create a group that can engage in creative discussions, without excluding variant perspectives. This is important, as while diametrically opposing perspectives may make the process more difficult, they are necessary if the specific function of scenarios, that of formulating possible futures (not necessarily probable) is to be achieved.
- In rural areas, especially in developing countries, the logistics of gathering people from a wider geographical area can be substantial. In the case of the Sesan River, for example, it proved difficult to arrange a meeting at the local level in the river basin because of difficult communications.
- The choice of factors to base the scenarios on is difficult; simplification is necessary while at the same time it involves a decrease of the complicated interdependency of issues in sustainable water management.
- As with all processes that involve stakeholder participation, it is necessary to define realistic outcomes at the very beginning of the process. The results of an exercise such as this with scenarios can increase the

understanding of the participants for common problems, can raise their awareness and knowledge, and can increase their capacity to define their own futures. However, it is necessary to make clear that the impact on policy made by local, regional and national authorities is completely dependent on those authorities' willingness to take on the results of the process.

Despite these reservations the use of stakeholder-formulated scenarios, especially in the Sesan and Tagus basins has demonstrated that the method can be a useful contribution to the development of sustainable policies for water management. The stakeholder participation process is further discussed in the following section.

## 14.4 STAKEHOLDER PARTICIPATION

Stakeholder participation and analysis is one of the most critical elements for practical IWRM implementation. In the STRIVER project, stakeholder participation was given an important place, as demonstrated through the series of stakeholder workshops that were conducted in all the four case basins each year. In all, 12 stakeholder workshops were conducted (3 within each basin), in addition to a range of targeted discussions with key stakeholders. During the first year, the workshops provided a platform for reviewing the initial stakeholder analyses and identifying the key stakeholders; introducing stakeholders to the project objectives and mapping stakeholder expectations, interests and problems; and fostering synergies with ongoing activities within each of the basins. In the second year, stakeholder workshops were used to steer research objectives and activities, including developing policy scenarios. The final year workshops were essential as a tool for collaboratively reviewing project outputs with stakeholders and the STRIVER team, as well as identifying avenues for further exploitation of STRIVER results.

The workshops not only helped in the integration of various perspectives of stakeholders from different sectors, but also different user groups with varied and conflicting interests. Experience showed that there was a relatively strong willingness among stakeholders to embrace the IWRM process irrespective of country, sector and/or occupational background; although the modalities remained fuzzy. The group dynamics observed at the workshops proved that it was possible to bring stakeholders that shared the waters for constructive dialogue, although the political, cultural and institutional context within each of the basins had a major impact on participation. Research-project focused stakeholder workshops proved to be a useful tool for enabling negotiations on transboundary management of water resources and identifying opportunities for resolving other water use conflicts. It was also noted that projects such as STRIVER can play a 'neutral role' in moderating the stakeholder workshops and motivating stakeholders with conflicting interests, by presenting research findings that were perceived to possess a strong air of legitimacy. The stakeholder workshops also played an important role in offering insights on IWRM practice from other basins around the world and thus promoting awareness and, to some extent, also capacity building. Ultimately, the stakeholder workshops helped in fostering linkages between the STRIVER researchers, managers, end users and policymakers, and at the same time improve acceptance of project outcomes. Other forms of stakeholder interactions was also part of the project (Figure 14.2).

**Figure 14.2** Interviews with local women about water flow fluctuations along the Sesan River in Cambodia. (Photograph: P. Stålnacke)

## 14.5 ENVIRONMENTAL FLOW METHODOLOGY WITH SCIENCE-STAKEHOLDER INTERACTION AND STAKEHOLDER PARTICIPATION

Hydropower regulation in rivers (turbine-type) began approximately 125 years ago and many of these regulations took place before the extent and importance of their ecological impacts was understood. Since the early 1970s these aspects have been given increased attention world-wide, both with respect to assessing the nature of the impacts as well as introducing abatement measures to reduce the negative impacts. These efforts at improvement have been aided by the introduction of concepts around minimum or environmental flow (EF) in the regulated stretches of the river. Halleraker and Harby (2006) in their review of EF methods define this as "a flow that as far as possible takes care of the entirety and integrity of the ecosystem, the different user interests, and the future resource base in the watercourse". EF is not only a concept but also a water management tool aimed at preserving species, habitats, livelihoods, and recreational values in water courses (King *et al.*, 1999).

The initial review of international literature conducted in STRIVER revealed more than 200 methods of assessing environmental flows, many of which were too complicated or demanding for use in developing countries. STRIVER therefore developed a methodology 'Pressure Impact Multi Criteria Environmental Flow Assessment' (PIMCEFA) that seeks to set water release rules that would ensure favorable water levels for river ecology and livelihoods, within the constraints of economic feasibility. This method has it basis in the operational Expert Panel method used in Norway for practical assessment of minimum water flows from hydro-power reservoirs.

The PIMCEFA method does not require comprehensive field investigations such as habitat and aquatic biological registrations and detailed river cross-section descriptions. According to this method, scientific inputs from river ecologists, hydrologists, environmental scientists, economists are first used to produce the basic knowledge about the functions, values and problems of the river basin. Then an expert panel is established with scientists, non-scientific experts and stakeholders: (i) define policy-relevant alternatives; (ii) identify river ecological and user interests (for which impacts are to be determined); (iii) draw optimal water level curves for each ecological value and user interest chosen; and (iv) discuss and develop pressure-impact curves for various flow regimes. Finally, the results are translated into a multi-criteria tool that pools the information; which can then be discussed with stakeholders in an accessible manner. The methodology itself requires a high level of collaboration with a range of stakeholders and the outcome of the work definitively integrates the knowledge of different expertise and the trade-off of contrasting interests. In the STRIVER project, the methodology was successfully developed in the Glomma and the Sesan river basins (Berge *et al.*, 2008). It was concluded that Environmental Flow has to be considered as one of the most important measures to reduce negative impacts from hydropower development. In the Sesan River, it is important for preserving aquatic productivity, biodiversity, as well as livelihoods for the local people in downstream reaches (Nesheim *et al.*, 2009).

## 14.6 TRANSBOUNDARY CONFLICTS AND WATER GOVERNANCE

Conflicts over water and the demands placed on water governance increase in transboundary settings, as in these contexts representatives of different national and sectoral interests from different countries must try to cooperate. In the national contexts, water laws and systems of administration are mostly unified (although competition between different agencies can and often does exist). In international rivers there is, with few exceptions, no unitary authority that can force actors in the water sector to comply with laws and agreements. Therefore, while water management regimes in transboundary rivers are especially dependent on efficient legal systems and communication, these systems often do not exist. This makes cooperation between the different organizations and institutions, such as governmental agencies and departments, more complicated. At the same time, NGOs and other stakeholder groups may find it hard to create efficient means of communication and influence with these diverse groups of policy-makers and managers (Gooch and Rieu-Clarke, 2008). Communication is a central issue in transboundary water regimes and takes place in a number of networks which consist of both official and unofficial actors. These networks include government officials, NGOs village elders, representatives for development banks etc., as well as infrastructures in the river basin such as hydroelectric power plant dams, and should be thought of as actor-networks, as networks of influences (Gooch, 2008a). Within these networks, the world-views and understanding of the actors and groups of actors strongly influences their treatment of information and knowledge. The scientific information provided by a project such as STRIVER may be accepted or rejected according to these mind-frames (Gooch, 2008b).

A key aspect of the work related to water governance was to develop a robust set of indicators for assessing governance within the context of transboundary IWRM. The entry point for such work was the recognition that, while considerable research has sought to measure governance in general, few initiatives have sought to tailor such initiatives to the specific context of (transboundary) IWRM. Having reviewed existing theory related to governance, IWRM and indicator analysis, STRIVER developed a set of indicator questions to examine the extent to which key aspects of good governance and IWRM have been embedded both within the applicable laws (international, regional and national), and the degree to which such laws have been implemented in practice. Collaboration with stakeholders, through the workshops and subsequent targeted interviews, constituted an essential element in ascertaining the extent to which the applicable laws had been implemented in practice. Stakeholder workshops also provide an important means for 'validating' the research developed by the STRIVER research team. Ultimately, the governance research concluded that IWRM implementation is heavily reliant on the existence of a broader governance framework that supports access to information and justice, as well as stakeholder participation in decision-making (Rieu-Clarke *et al.*, 2008).

## 14.7 POLLUTION MODELING WITH STAKEHOLDERS

In STRIVER water pollution related to phosphorus (P) and nitrogen (N) losses from agriculture was addressed. Accurate quantification is a major challenge facing the research community as managers and policy makers increasingly rely on outputs from river basin models when evaluating environmental changes and management actions. In a science-policy context this implies both selecting the appropriate tools and making sure that the management scenarios have actual relevance.

Pollution modeling was carried out in Glomma (sub-basins Hunnselva and Lena) and Tungabhadra using the same methodological approach, i.e., implementing the same river basin modeling tool(s) and involving local stakeholders at different stages of the modeling process, including scenario development and analysis. The overall objective was to obtain information about the suitability of models for the assessment of nutrient losses under different management scenarios in order to meet surface water quality targets and analyze the possibilities for mutual transfer of know-how (e.g., experiences, concepts, results) and technology (e.g., methodologies, models) in the twinned basins based on existing data and procedures. In order for models to be useful tools in IWRM, it is paramount that the managers can trust the results of the modeling, i.e. that sound scientific advice and economically defensible decisions can be made based on the results. The quantification tools should aim to be accurate and responsive to changes in land use and land management. A general guideline is to choose a model in compliance with the data available. In this respect, local knowledge about the physical conditions and management practices in the basin, in addition to having an ongoing dialogue with stakeholders, is a prerequisite in obtaining reliable scenario results.

Two models were applied, i.e. the TEOTIL (Tjomsland *et al.*, 2009) and SWAT (Neitsch *et al.*, 2002) respectively. SWAT relies on detailed spatial input data, among others soils, cropping systems and climatological data, while less information are needed for the TEOTIL model, which operates mainly on the basis of so-called export coefficients for different land use types. The overall modeling performance was found to be reasonable; however, shortages and gaps in the required data were identified during the modeling process leading to assumptions and data constructions. Specific modeling results and management recommendations for the Glomma and Tungabhadra river basins are reported in Barkved *et al.* (2008) and Lo Porto *et al.* (2008) respectively.

Participatory modeling approaches include involving stakeholders in model exercises, thereby incorporating local knowledge and understanding of the natural system and ensuring local participation in decision making. In such a process it is important to identify and understand the values and motives of a wide range of stakeholders involved in the decision making process and in this respect interaction with stakeholders is important. The STRIVER project showed that interaction and co-operation between scientist and local stakeholders of the respective basins in the development and implementation of scenario modeling is a prerequisite in IWRM (Grizzetti *et al.*, 2008). In the case of the Glomma sub-basins Hunnselva and Lena, measures related to agricultural practices to comply with environmental policies were selected and included among others conservative tillage and optimal fertilization application. Hunnselva is part of Phase 1 in the implementation of the EU's Water Framework Directive (WDF) and as such, the modeling results have been used as supporting knowledge in the planning of mitigation measures in the basin. In the Tungabhadra river basin, the stakeholders were interested in climate change impacts, improved sewage treatment, irrigation technology and changed rice production.

A key message in pollution modeling with stakeholder involvement is that knowledge about the local conditions is an important asset to ensure reliable results. The experiences also showed that stakeholder involvement at different phases of the modeling process, such as model input preparation, scenario building and discussions of modeling outcome, plays a key role.

## 14.8 LAND AND WATER USE INTERACTIONS

The analysis of land and water use interactions and its implications for IWRM was based on a comparative study in two river basins, the Tungabhadra and Tagus. The relationship between land use and water resources is twofold: on one hand, land use exerts a major control on water use; but it also has great potential for modifying the hydrological cycle within the river basin, thus affecting the quantity and quality of water resources. As several studies have pointed out, the effects of land use change on the water cycle can have a magnitude equivalent to those of climate change. The interaction between land use and the water cycle is relatively well known, although it is often ignored in water resources management plans.

The analysis confirmed that there has been significant land use change in the last decades in both the Tagus and the Tungabhadra basins (Beguería *et al.*, 2008), which have been accompanied by new water demands and resultant conflicts between users. In the case of the Tagus, the increase of urban land cover has been the most significant land use change in the Spanish part of the basin; whereas in Portugal the increase of natural vegetation cover in the headwaters has been predominant (CORINE Land Cover 1985/1990 and 2000). In the Tungabhadra basin, the most significant change has been from predominantly rain-fed to irrigated agriculture, which has had a large impact on water use in the basin. An increase in the area covered by forest has also occurred in the headwater areas, as reported by municipal statistics and by a satellite-imagery survey. At the same time, urbanization has increased in both basins. Unfortunately, while these changes have occurred, there are still few institutional arrangements at the basin level to plan and manage water resources in an integrated manner. These would need to take into account the consequences of land use change on the quantity and quality of water resources and the increasing water demands and conflicts between various competing sectors and at various levels – macro as well as micro level of distributaries, communities and individual users. IWRM appears as the ideal framework to help in resolving conflicts arising from competing land uses. However, transboundary conflicts over water sharing in both basins are an impediment to the implementation of IWRM.

The STRIVER study strongly recommends improving water use efficiency in the basins in order to address the water scarcity and water management problems within the agriculture sectors. Measures such as new methods of cultivation like the System of Rice Intensification, introducing new crop varieties suited to water stress conditions, reducing evapotranspiration and transportation losses, improving fisheries and other non consumptive water uses can help not only in improving water use efficiency, but also in addressing livelihoods issues and the sustainability of agro-ecosystems. There are fragmented efforts at developing technological and institutional innovations in the basin that need to be synthesized and brought together as part of an integrated strategy. For example, Participatory Irrigation Management (PIM) in Tungabhadra has been initiated to involve farmers in water management. Similarly, water pricing has been introduced to rationalize water use. But, in practice, PIM has shown poor results due to lack of proper co-ordination, lack of willingness to collect water charges and inadequate legal and institutional framework.

Integrating water management at different levels could address various problems and at the same time improve water use efficiency (WUE). This requires strong institutional and policy support and political will for implementation. Although the water policies in both cases demonstrate the willingness, the implementation is poor, due in part to a number of institutional constraints and lack of capacity. Capacity building is needed to prepare water users, water managers and policy makers for adopting an integrated approach to water management.

## 14.9 STRIVER GUIDANCE AND STRATEGY DEVELOPMENT

Strengthening science-policy linkages was given a major emphasis in the STRIVER project. This has resulted in a series of policy and technical briefs that were produced in order to communicate the main technical and policy messages to the stakeholders. The content of the briefs has been presented in a form that the stakeholders should find easy to understand. Attempts were also made to present the briefs to managers and policy makers in various meetings and workshops and on a one-to-one basis. Some of the briefs were translated into local languages to benefit the end users. The main objective was to disseminate the project results to the most relevant stakeholders. In total, 22 policy briefs have been published (Table 14.1).

The 13 technical briefs were specifically addressed to managers and scientific community in the case basins. The major focus was on synthesizing technical results from a number of issues (Table 14.2).

The basin-specific briefs have been presented at the stakeholder meetings and also disseminated at larger events such as the World Water Forum 5 in 2009. The effectiveness of the briefs is yet to be monitored, but it is likely that such briefs can better target end-users compared to ordinary research articles and lengthy reports.

**Table 14.1** Issues covered in 22 STRIVER Policy Briefs.

| | |
|---|---|
| PB1. | Livelihoods: Fisheries in Tungabhadra sub-basin, India. |
| PB2. | Water Regimes in Transboundary Heavily Modified Rivers. |
| PB3. | Water valuation and pricing – when are they useful in water management? |
| PB4. | IWRM and Traditional systems: Tanks in the Tungabhadra system. |
| PB5. | Competing Water Uses in Tungabhadra Sub-basin, India. |
| PB6. | Communication in IWRM in Transboundary Rivers. |
| PB7. | Modelling water and nutrients balance in Tungabhadra river basin: scenario analysis and management recommendations. |
| PB8. | Assessing governance in the context of IWRM. |
| PB9. | Environmental Flows (EF) in IWRM – with reference to the hydropower regulated Glomma River in Norway and Sesan River in Vietnam/Cambodia. |
| PB10. | Modelling Water Pollution with Stakeholders involvement. The twinned experience of Glomma (Norway) and Tungabhadra (India) river basins. |
| PB11. | Modelling nutrients in the Glomma river basin: scenarios and management recommendations. |
| PB12. | Environmental flow, biodiversity and livelihoods – the causal chain. |
| PB13. | Strategies and Recommendations for River Basin Management in the Sesan. |
| PB14. | A comparative assessment of IWRM principles in Tagus (Spain and Portugal). |
| PB15. | Strategies and Recommendations towards an IWRM approach in Tungabhadra sub-basin, India. |
| PB16. | Strategies and recommendations for IWRM in Glomma. |
| PB17. | Participatory training in canal irrigation in Andhra Pradesh: The JalaSpandana Experience. |
| PB18. | Willingness to pay – unwillingness to be paid. The politics of water pricing in canal irrigation in Andhra Pradesh, India. |
| PB19. | The Tagus River Basin – Spain and Portugal. |
| PB20. | Capacity Building Activities in the STRIVER Case Basins. |
| PB21. | Stakeholder Participation in the STRIVER Project. |
| PB22. | IWRM Recommendations and strategies. |

Source: http://kvina.niva.no/striver/Disseminationofresults/STRIVERTechnicalBriefs/tabid/78/Default.aspx

**Table 14.2** Issues covered in 13 STRIVER Technical Briefs.

| | |
|---|---|
| TB1. | Land use change and water resources in the Tagus and Tungabhadra basins. |
| TB2. | Actor-Network Theory in Water Management – a help or hinder for understanding water management regimes? |
| TB3. | Alternative approaches to valuing irrigation water. |
| TB4. | Using choice experiments to value irrigation water. |
| TB5. | Long-term flood records for flood risk management. |
| TB6. | Pressure-Impact Multi-Criteria Environmental Flow Analysis in the Glomma River. |
| TB7. | Multi-criteria environmental flow analysis in the Sesan River. |
| TB8. | BartonMiljø- og Brukerbasert Vannføring: multikriterie- metode med konsekvenskurver. (In Norwegian) |
| TB9. | A case of transboundary water agreement – the Albufeira Convention. |
| TB10. | Development in river basin management in Portugal – past and future perspectives. |
| TB11. | IWRM status in the Tagus basin, Spain. |
| TB12. | A limnological study in the Sesan River in Cambodia during the dry season: focus on toxic cyanobacteria and coliform bacteria. |
| TB13. | Comparative assessment of IWRM methods. |

Source: http://kvina.niva.no/striver/Disseminationofresults/STRIVERTechnicalBriefs/tabid/79/Default.aspx

### 14.9.1 Major findings

Overall, STRIVER has so far demonstrated that:

- Stakeholder participation is one of the most critical elements for practical IWRM implementation. It helps not only in the integration of various perspectives of stakeholders from different sectors, but also of different user groups with varied and conflicting interests.
- Research projects such as STRIVER can act as an independent facilitator, and provide a neutral platform for stakeholder dialogue, which ultimately can facilitate the IWRM process. There was relatively a strong willingness among stakeholders to participate in the IWRM process during the workshops, irrespective of their country, sector and occupational background. The stakeholder workshops could also play an important role in capacity development initiatives.

- Cross-border cooperation is a difficult process and sensitive process in all the three transboundary river basins and can be a hindrance to IWRM if not addressed. It was only in the stakeholder workshops STRIVER could bring together actors from different countries or provinces sharing the river waters.
- The development and show-cases of various 'tools' (e.g., environmental flow, pollution models, water pricing, Actor-Network analyses and scenarios) is of high interest and regarded of high value for the water managers. One common feature with all the applied and tested tools where that they promote dialogue and integration between different actors of IWRM as well as between scientists and stakeholders.
- There exists a considerable number of 'IWRM-initiatives', IWRM-plans and policies in all the four basins, but practical IWRM implementation is generally lacking, except in the case of Glomma.

For the individual case basins the following was particularly noted:

- considering livelihood issues for all sectoral and marginal groups could increase the water use efficiency and facilitate the move towards IWRM (as seen in Tungabhadra case);
- the introduction of water pricing as a regulatory measure could help in increasing water use efficiency, provided there is a proper institutional framework in place for implementation at a basin or sub-basin level (Tungabhadra and Sesan cases)
- implementation of a simple environmental flow method for hydropower water release could take into joint consideration of all sectoral needs and demands (Sesan case);
- quantifying pressure and water quality impacts with pollution models was regarded of high value by water managers in the implementation of the EU Water Framework Directive (Glomma case);
- Unregulated land use change will have a negative impact on water use and management, if not integrated with overall development in the basin (as seen in Tungabhadra and Tagus).

### 14.9.2 Concluding remarks: some reflections on how to strengthen European research from the perspective of IWRM

Compared to many other R&D projects, twinning projects such as STRIVER have the flexibility to interact with water managers and policy makers. However, this requires that scientists take the role of 'scientific ambassadors'. In addition, it also demands more time and resources in terms of logistics required for travels, meetings and discussion forums. In some cases, managers and policy-makers can be initially skeptical about 'spending' their valuable time with researchers, something that was also observed in the initial stages of STRIVER. In the course of time, STRIVER researchers managed to establish trust and confidence amongst the stakeholders. For researchers, however, stakeholder interaction may come at the expense of scientific publications and career advancement. If results from projects such as STRIVER are to reach managers and policy makers, it is worthwhile to spend the resources and establish formal links with stakeholders (from local level, to high-level water managers and policy makers), just as it is with the research community particularly related to the implementation of IWRM or sustainability programs. This should be targeted throughout the whole course of the project. Such links can constitute a win-win situation for both researchers and managers, and also help improve understanding of the research problems to be addressed, including jointly developing appropriate tools, scenarios and policy guidelines. STRIVER researchers could also observe that stakeholders were interested in capacity building that could be formalized in the projects, this was seen at first-hand during the field trips and workshops. In more practical terms, it is recommended that funding agencies like DG Research at the EC continue with the twinning mechanisms with more emphasis on twinning between the scientific community and stakeholders and within a timeframe of at least 4–5 years project duration.

## 14.10 REFERENCES


Barkved, L.J., Deelstra, J., Grizzetti, B. and Bouraoui, F. (2008) Modelling nutrients in the Glomma river basin: scenarios & management recommendations. STRIVER Policy Brief No. 11. (http://www.striver.no/diss_res/files/STRIVER_PB11.pdf).

Barton, D.N., Raju, K.V. and McNeill, D. (2008a) Water valuation and pricing – when are they useful in water management? STRIVER Policy Brief No. 3. (http://www.striver.no/diss_res/files/STRIVER_PB3.pdf)

Beguería, S. Vicente-Serrano, S., López-Moreno, I., Lana-Renault, N. and García-Ruiz, J.M. (2008) Land use change and water resources in the Tagus and Tungabhadra basins. STRIVER Technical Brief No. 1.

Benito, G., Botero, B., Lana-Renault, N., García-Ruiz, J. M., Vicente-Serrano, S. and Beguería, S. (2008) Long-term flood records for flood risk management. STRIVER Technical Brief No. 5.

Berge, D., Berge, D., Nhung, D.K., Hoang, P.T.T., Barton, D.N. and Nesheim, I. (2008) The use of Environmental Flow in IWRM, with reference to the hydropower regulated Glomma River in Norway and Sesan River in Vietnam/Cambodia., STRIVER Report No. D 8.1.

CORINE Land Cover 1985/1990 and 2000. http://www.eea.europa.eu/publications/tech40add

Gooch, G.D. and Stålnacke P. (Eds.) (2006) Integrated Transboundary Water Management in Theory and Practice: Experiences from the New EU Eastern border. London: IWA Publishing

Gooch, G.D. (2008a) Actor-Network Theory in water management – a help or hinder for understanding water management regimes? STRIVER Technical Brief No. 2.

Gooch, G.D. (2008b) Communication in IWRM in transboundary rivers. STRIVER Policy Brief No. 6. (http://www.striver.no/diss_res/files/STRIVER_PB6.pdf).

Gooch, G.D. and Rieu-Clarke, A. (2008) Water regimes in transboundary heavily modified rivers. STRIVER Policy Brief No. 2. (http://www.striver.no/diss_res/files/STRIVER_PB2.pdf)

Gooch, G.D., Allan, A., Rieu-Clarke, A. and Baggett, S. (2010) The Science-Policy-Stakeholder Interface in Sustainable Water Management. Creating interactive, participatory scenarios together with stakeholders, in Gooch, G.D. and P. Stålnacke 'Science, Policy and Stakeholders in Water Management' London: Earthscan.

Grizzetti, B., Lo Porto, A., Barkved, L.J., Bouraoui, F., Deelstra, J. and Joy, K.J. (2008) Modelling water pollution with stakeholders involvement – the twinned experience of Glomma (Norway) and Tungabhadra (India) river basins. STRIVER Policy Brief No. 10. (http://www.striver.no/diss_res/files/STRIVER_PB10.pdf)

Halleraker, J.H. and Harby, A. (2006) International methods for deciding environmental flow – which of these are applicable in Norway? NVE Miljøbasert vannføring, Report 9 2006.

King, J. Tharme, R.E. and Brown, C. (1999) Definition and implementation of instream flows., World Commission on Dams, Thematic Report.

Lo Porto, A., Barkved, L.J. and Gosain, K.A. (2008) Modelling water and nutrients balance in Tungabhadra river basin: scenario analysis and management recommendations. STRIVER Policy Brief No. 7. (http://www.striver.no/diss_res/files/STRIVER_PB7.pdf).

Neitsch, S.L., Arnold, J.G., Kiniry, J.R., Williams, J.R. and King, K.W. (2002) Soil Water Assessment Tool Theoretical Documentation. Grassland, soil and water research laboratory. Agricultural Research Service. Blackland Research Center. Texas Agricultural Experimental Station. Temple Texas, 76502.

Nesheim, I., Berge, D., Nhung, D.K. and Gooch, G.D. (2009) Environmental flow, biodiversity and livelihoods – the causal chain. STRIVER Policy Brief No. 12. http://kvina.niva.no/striver/Disseminationofresults/tabid/70/Default.aspx

Nesheim, I., McNeill, D., Stålnacke, P., Sekhar, N.U., Grizzetti. B., Allen, A.A., Barton, D., Beguería-Portugés, S., Berge, D., Bouraoui, F., Campbell, D., Deelstra, J., García-Ruiz, J.M., Gooch, G.D., Joy. K., Lana-Renault, N., Lo Porto, A., Machado M., Manasi S., Nhung D.K., Paranjape S., Portela M.M., Rieu-Clarke A., Saravanan V.S., Thaulow, H. and Vicente-Serrano, S. (2008) First IWRM assessment report for the four case basins: Glomma, Tagus, Sesan and Tungabhadra. STRIVER Report D5.1.

Rieu-Clarke, A., Allan, A. and Magsig, B.O. (2008) Assessing governance in the context of IWRM. STRIVER Policy Brief No. 8. (http://www.striver.no/diss_res/files/STRIVER_PB8.pdf).

Tjomsland, T., Selvik, J.R. and Brænden, R. (2009) Teotil. Model for calculation of source dependent loads in river basins. NIVA, Oslo.

IWA Publishing's authorised EU representative for General Product Safety Regulations is Diane D'Arras, 15 rue Duret, 75116 Paris, France, e-mail: safety@iwap.co.uk.

Printed and bound by CPI Group (UK) Ltd, Croydon, CR0 4YY
12/05/2026
02108874-0019